Reliability and Physics-of-Healthy in Mechatronics

Reliability of Multiphysical Systems Set

coordinated by
Abdelkhalak El Hami

Volume 15

Reliability and Physics-of-Healthy in Mechatronics

Edited by

Abdelkhalak El Hami
David Delaux
Henri Grzeskowiak

iSTE

WILEY

First published 2022 in Great Britain and the United States by ISTE Ltd and John Wiley & Sons, Inc.

ISTE Ltd
27-37 St George's Road
London SW19 4EU
UK

www.iste.co.uk

John Wiley & Sons, Inc.
111 River Street
Hoboken, NJ 07030
USA

www.wiley.com

Library of Congress Control Number: 2022941493

British Library Cataloguing-in-Publication Data
A CIP record for this book is available from the British Library
ISBN 978-1-78630-881-8

Contents

Preface

The emergence of Covid-19 in March 2020 was a turning point throughout the industrial world This situation has highlighted some weaknesses in our industries, pushing the paradigm of development, competition and development time reduction.

More than ever, we have to be not only agile in our business but also without doubt to secure the usage of our products. It is very hard to combine variability and stability, acceleration and standardization. More than ever, the world is pushing all industrics to the limit of reliability engineering.

Truth be told, reliability is a beautiful marketing word, which is used many times a day! But behind this word, there is a beautiful concept of pragmatic actions composed of analysis, investigations and talented people.

What kind of approach can designers, engineers and managers apply?

Whatever the business (aeronautic, space, defense, automotive, electronic consumer, etc.), how can we validate a design quickly and safely?

This book illustrates simply, but with many details, the state of the art of reliability science. The authors wish to expose clear reliability disciplines and applications through concrete examples from their industries and real life, based on industrial experience.

Many experts are convinced that reliability involves not only statistics but also multidisciplinary sciences, which combine materials, tests, simulations, quality tools, manufacturing, electronics, mechatronics, environmental engineering, Big Data science, etc.

That is why, in this innovative book, the reader will discover an incredible study on some concepts of a predictive reliability model of microelectronics from automotive, aeronautic and space missions based on entropy and Physics-of-Healthy (Chapters 1 and 2).

For a complex mechatronic system in aircraft, failure risks have to be identified in the early phase of the design.

In the automotive or aeronautic industries, the simulation of fatigue is used widely and efficiently. The problem is the variability of inputs such as fatigue parameters and life curves. The probabilistic fatigue and reliability simulation discussed in Chapter 3 will be very interesting in this regard.

A large number of complex materials such as 3D printing, composites and thermoplastics are used to optimize the reliability performance of a system. However, the difficulty is usually linked to the lifetime currently being between 10 and 20 years. How can we assess reliability performance under temperature stress and represent a complete mission profile? Of course, Arrhenius is a frequently used model; however, it can also be "erroneous" due to the variability in the energy of activation. Therefore, a new mode probabilistic model is put forward in Chapter 5.

Yes, there is a "before" and "after" Covid-19. But the challenge of reliability in our industry is more than ever a burning topic. We hope that readers will find some tips that can efficiently help all designers to make strong decisions.

The authors would like to dedicate this book to all the people who have lost their loved ones during the Covid-19 pandemic around the world.

Abdelkhalak EL HAMI
David DELAUX
Henri GRZESKOWIAK
July 2022

List of Acronyms

BAZ	Boltzmann–Arrhenius–Zhurkov
BEOL	Back End Of Line
BTI	Bias Temperature Instability (NBTI or PBTI)
CA	Constructional Analysis
CCC (or CC)	Channel Cold Carrier
CDF	Cumulative Distribution Function
CFET	Complementary Field Effect Transistor
CHC (or HC)	Channel Hot Carrier
$c(x_i)$	Energy factor related to stress parameter x_i
DIBL	Drain-Induced Barrier Lowering
DoE	Design of Experiment
DSM	Deep-SubMicron technology
E_a	Activation energy

$E_{a\_effective}$ or E_{aa}	Effective activation energy related to multiple stress reliability test sequence
ECSS	European Cooperation on Space Standardization
EES	Electron–Electron Scattering
EM	ElectroMigration
EMC	ElectroMagnetic Compatibility
EOS	Electrical OverStress
EOT	Equivalent Oxide Thickness
ESD	ElectroStatic Discharge
EVD	Extreme Value Distribution
FEOL	Front End Of Line
FF	FinFET
FI	Fan In
FO	Fan Out
FRAME	Failure Risk Analysis MEthodology
G	Gibbs free energy
GaAs	Gallium Arsenide
GaN	Gallium Nitride
GIDL	Gate-Induced Drain Leakage
HBD	Hard BreakDown
HCD	Hot Carrier Degradation

HCI	Hot Carrier Injection mechanism
I_{BO}	Current burnout limit value
IFR or $\lambda(t)$	Instantaneous Failure Rate or hazard function
I_g	Peak gate current during stress
I_{G_leak}	Gate leakage current between gate and source at V_{GS} and V_{DS}
I_{G_LBO}	Gate leakage current between gate and source at high V_{GS} and V_{DS} close to burnout or breakdown
I_{HCG}	Current due to hot carrier generation
ILD	Inter-Layer Dielectric
I_{sub}	Peak substrate current during stress
k	Boltzmann's constant (1.3807·10^{-23} J/K or 8.6174 10^{-5} eV/K)
M-STORM	Multi-phySics mulTi-stressOrs predictive Reliability Model
M-TOL	Multiple Temperature Overstress Life test
MESFET	MEtal–Semiconductor Field-Effect Transistor
MR	Maximum Rating
MVE	Multi-Vibrational Excitation
NAF	NMOS Acceleration time Factor
NBTI	Negative Bias Temperature Instability
NTF	DC NMOS acceleration time Factor
OCM	Original Component Manufacturer
OEM	Original Equipment Manufacturer
PAF	PMOS Acceleration Factor

P_{BO}	Power burnout limit value
PBTI	Positive Bias Temperature Instability
PDF	Probability Density Function
PDfR	Probabilistic Design-for-Reliability
PHEMT	Pseudomorphic High Electron Mobility Transistor
PMOS	P-type Metal Oxide Semiconductor
PoF	Physics of Failure
PoH	Physics-of-Healthy
PoS	Physics of Stress
P_q	Poisson statistic function
q	Number of failures in Poisson statistic function
QBD	Charge at breakdown
SAC	Self-Aligned Process
SBD	Soft BreakDown
S_e	Electrical indicator or signature of the failure mechanism
S_i	Stress parameter for i as current I, voltage V or P_{DC} power consumption or P_{in} signal input power
S_{iBO}	Stress parameter at burnout limit value
SiC	Silicon Carbide
SM/SV	Stress Migration/Voiding
SOA	Safe Operating Area
SVE	Single Vibrational Excitation

TDDB	Time-Dependent Dielectric Breakdown
TSRM	Transition State Reliability Model
TST	Transition State Theory
γ_i	Stress factor as the percentage of burnout limits for label i as current I, voltage V, power P_{DC}, signal input power P_{in}
$\in$ or $\in(t)$	Gibbs free energy for a device

Part 1

Entropy and Physics-of-Healthy: Some Concepts to Model Predictive Reliability of Microelectronics for Automotive, Aeronautic and Space Missions

Introduction to Part 1

Electronic components and devices, including equipment for systems, are fabricated from materials and structures that degrade over time under normal operational conditions. It is therefore necessary to anticipate and quantify the occurrence of system failures, but the goal of "reliability engineering" is to first clarify the failure paradigm in terms of statistics and physics of failure (PoF) or simply the parts count approach. Statistical approaches and underlying mathematics have been developed over the last century to describe failure rates, including the well-known bathtub curve showing a schematic of failure rate behavior with time.

According to Wikipedia, predictive (or logical) analysis encompasses a variety of techniques derived from statistics, data extraction and game theory that analyze past and present facts to make predictive assumptions about future events.

In the first part of this book, we share our perception of the experimental development applied to existing reliability and maintenance paradigms that has taken place over several decades, been presented by many authors and appeared in numerous papers.

Chapter 1 aims to i) establish how applied engineering for failure prediction and SHM (structural health monitoring) of electronic equipment and systems is implemented, and ii) present basic statistical tools defined for reliability modeling implementation as well as studies related to active microelectronic parts. Chapter 2 focuses on experimental applications and addresses how innovative technologies and commercial-off-the-shelf (COTS) devices are considered.

Introduction to Part 1 written by Alain BENSOUSSAN, Joseph B. BERNSTEIN and Alain BRAVAIX.

How do we implement failure prediction and SHM of electronic equipment and systems?

The high-temperature operating life test (HTOL), performed according to the JEDEC JESD22-A108 standard, is relatively poor for predicting mean-time-to-failure (MTTF) and the associated constant failure rate; it has been shown to be largely ineffective for wearout modeling as no failures are observed. Furthermore, the MTTF (50%) is not a zero-failure guarantee (rather, by definition, it is a 50% lot failure) and we should instead define an MTTF of 0.1% for high reliability application. At the same time, attention to long-life cycles on the part of manufacturers has nearly vanished due to the relatively overwhelming demand in the consumer market sector.

Three key questions are considered:

i) How do we apply reliability prediction tools to innovative technologies?

ii) How are hypotheses constructed and methodology implemented in these tools, and to what extent are they able to cover emerging technologies and applications?

iii) For the prognostic failure model (PFM), what could a proposed guideline be to organize all concepts?

The aim of reliability engineering is to anticipate and quantify failure occurrences but the science required first clarifies the paradigm in terms of statistics and the physics of failure (National Academy of Sciences 2008; White and Bernstein 2008) or the parts-count failure approach (DoD 1991; Lall et al. 1997; JEDEC 2011). Many questions are raised which we attempt to address, listed as follows.

In terms of physics and phenomenological degradation

– What are the existing standards, their advantages and drawbacks?

– How can we manage random and wearout failure rate models to implement predictive reliability and maintenance?

– What role do “stressors” play?

– How are “indicators” characterized and which criteria value should be defined?

– How do we deal with multiple stresses?

– In very complex new technologies (less than 7 nm, for example, FinFET, GaN, Carbone NanoTube, etc.), what are the new failure mechanisms?

– What if multiple failure mechanisms coexist (JEDEC 2011)?

– What is the impact on the accelerating factor (AF) in such complex mixed models (multiple stresses, multiple failure mechanisms) (Bernstein et al. 2014)?

– How do we apply reliability prediction tools to innovative technologies?

– What are the hypotheses made of and methodology implemented in these tools, and to what extent are they able to cover emerging technologies and applications?

– How to consider the effect of initial device quality, the influence of use and design options, the effect of mission profiles to model the robustness (Robustness Group ZVEI 2013) and the reliability of complex systems and new technologies (T.S.Q. &. R. Publications n.d.; Berthon et al. 2014; QUANTERION 2015)?

In terms of applications and experiments

In very complex new technologies, observed failure mechanisms can often be accelerated simultaneously, causing a dilemma for reliability prediction.

In the domain of telecommunications, automotive, aerospace, satellite, military and the like, the need for accurate reliability prediction is as important as it has ever been. At the same time, attention to long-life cycles on the part of manufacturers has nearly vanished due to the relatively overwhelming demand in the consumer market sector:

– How to explore field return and define optimized maintenance survey based on reliability prediction tools?

– Assuming a qualification test experiment on 130 samples tested during 2000 hours of 150°C HTOL with 0 failure observed.

– What would be the device operating projected lifetime verified if the operational application is at T = 45°C?

– What should be the sample size and duration to demonstrate a cumulative density function (CDF) equal to less than 0.1% after 30 years in mission operation?

– For the purpose of reliability prediction, do we need to consider time-to-failure for each failure mechanism either for a constant failure rate based on a Poisson statistic, or for wearout failure mechanisms?

– What would be the equivalent total λ_{total} and corresponding time-to-failure if all failure mechanisms are equally activated?

– What are the basic mathematics?

Thermodynamics modeling established by Boltzmann (1886), Arrhenius (1896), then Evans and Polanyi (1938), Wigner (1938), Glasstone et al. (1941), Hammond (1955), Drenick (1960), Snow (1965), Sedyakin (1966), Cox (1972) and periodically improved by Eyring et al. (1980), McPherson and Baglee (1985), McPherson (2019), or recently by Suhir (2010, 2013). Modeling was supported by a series of experimental papers from MIT (Jin and Del Alamo 2012), RAC University of Maryland, University of Padua and Meneghesso's team (Meneghesso et al. 2010), INTEL and IBM teams (Stathis et al. 2018). Since the breaking of 217+ standards, FIDES French group guidelines are still in place and its next evolution is thought to consider a non-constant failure rate model for predictive reliability. Not leaving aside main actors in this paradigm development, we have to name the most representative experts in the field which have influenced the writing of these chapters: J. Stathis, V. Huard, A. Bravaix, J. Bernstein, M. Meneghini, H. Zanoni, E. Wu and J. Sune, P.A. Tobias and D.C. Trindade, M. Nikulin and V. Couallier.

The paradigm of the transition state theory (TST) developed by Wigner (1938) and by Evans and Polanyi (1938) is an approach we can also benefit from by adapting the concept of a unified semiconductor reliability model and multiple failure mechanisms related to physics of degradation. Indeed, early in the last century, the TST was applied to chemistry transformations by Eyring et al. (1980) and Glasstone et al. (1941). The TST was developed through chemistry based on Hammond's postulate (Hammond 1955) applied to physical organic chemistry.

Multiple failure mechanisms and physics of degradation in semiconductors may occur in a single set of time-to-failure data but without obvious points of inflection to help separate the mechanisms. McPherson (McPherson and Baglee 1985; McPherson 2019) provides the basics of reliability modeling by stating this.

The Gibbs free energy diagram, recalled by McPherson (2019), provides the main inputs for describing the effect of multiple environmental stressors, including the entangled accelerating factor which is fully detailed in Chapter 1. Numerical applications for a case study on FinFET technology, assuming three major failure mechanisms defined by the following Arrhenius reliability parameters, are also detailed.

These approaches in mathematics and physics show how the activation energy is an Eyring model related to the stress and temperature applied and can no longer be considered as a constant to extrapolate an experiment under high stress to nominal mission operation profile. This is why several end user industries and institutions are very cautious to perform life test conditions as close as the nominal conditions because of the change of activation energy sometimes wrongly attributed to new (or different) failure mechanisms while it is simply explained by the interaction of stress and the effect of temperature on the measured activation energy or, to a great extent, to the Eyring law.

From these equations, we can observe that the equivalent activation energy is dependent on the temperature T0 and increases with temperature. So, under the stress conditions, the failure mechanism model is a non-uniform acceleration mechanism.

The aim of Chapter 1 is to present basic statistic tools defined for reliability modeling implementation and studies related to active microelectronic parts (integrated circuits, power transistors, etc.) when used in an operational environment for long-term high reliability applications.

Basic mathematics on series-parallel system reliability are presented with some approximation considerations for distribution queue statistics.

These conclusions and perspectives open the door for Chapter 2, which is related to predictive reliability supported by experimental and physics of failure (PoF) or what we call Physics-of-Healthy (PoH).

Physics of degradation in semiconductors may occur in a single set of time-to-failure data but without obvious points of inflection to help separate the mechanisms. McPherson (McPherson and Baglee 1985; McPherson 2019) stated that:

> [...] generally, materials/devices exist in metastable states. These states are referred to as being metastable because they are only apparently stable. Metastable states will change/degrade with time. The rate of degradation of the materials (and eventual time-to-failure for the device) can be accelerated by an elevated stress (e.g. mechanical stress, electrical stress, electrochemical stress, etc.) and/or elevated temperature.

The aim of Chapter 2 is to concentrate on experiments supported by models and field returns showing the "true life" observing simultaneous stress environment (various mission profiles) and multiple failure mechanisms.

References

Arrhenius, S. (1896). Arrhenius, Ueber den Einfluss des Atmosphärischen Kohlensäurengehalts auf die Temperatur der Erdoberfläche. *Proceedings of the Royal Swedish Academy of Science*, 26(1), 6–11.

BELLCORE (2011). SR-332, Reliability Prediction Procedure for Electronic Equipment. BELLCORE Telcordia.

Bensoussan, A., Durier, A., Bernstein, J.B. (2017). M-Storm: A new approach to model electronics components reliability. *Automotive Power Electronics (APE 2017)*, Paris.

Bernstein, J.B. (2015). Reliability prediction for aerospace electronics. *IEEE Aerospace Conference*, Big Sky, MN, April.

Bernstein, J.B., Gabbay, M., Delly, O. (2014). Reliability matrix solution to multiple mechanism prediction. *Microelectronics Reliability Journal*, 54, 2951–2955.

Berthon, J., Regis, D., Hubert, G. (2014). Utilisation des composants DSM dans le contexte aéronautique. *19ème Congrés de Maitrise des Risques et Sûreté de Fonctionnement*, Dijon, 21–23 October.

Boltzmann, L. (1886). Boltzmann, the second law of thermodynamics. Populare Schriften. *Essay 3, Address to a Formal Meeting of the Imperial Academy of Science*, 28 May.

Bravaix, A. (2019). Part 2 – DDR3 FF20 Reliability Analysis – ISEN. Internal report, IRT- ISEN, Toulouse, September.

Cox, D. (1972). Regression models and life-tables. *Journal of the Royal Statistical Society. Series B (Methodological)*, 34(2), 187–220.

Creveling, C. (1956). Increasing the reliability of electronic equipment by the use of redundant circuits. *Proc. IRE*, 44, 509–515.

Dhillon, B. and Rayapati, S. (1986). A method to evaluate reliability of three-state device network. *Microelectronics Reliability Journal*, 26, 525–554.

DoD (1991). MIL-HDBK-217, Military Handbook for Reliability Prediction of Electronic Equipment. Washington, DC, December.

Drenick, R. (1960). Mathematical aspects of the reliability problem. *Journal of the Society for Industrial and Applied Mathematics*, 8(11), 125–149.

Evans, M. and Polanyi, M. (1938). Inertia and driving force of chemical reaction. *Faraday Society (London) Trans.*, 34, 11–29.

Eyring, H., Lin, S., Lin, S. (1980). *Basic Chemical Kinetics*. John Wiley & Sons, New York.

Glasstone, S., Laider, K.J., Eyring, H. (1941). *The Theory of Rate Processs: The Kinetics of Chemical Reaction, Viscosity, Diffusion and Electrochemical Phenomena*. McGraw-Hill, New York.

Hammond, G. (1955). A correlation of reaction rates. *Journal of the American Chemical Society*, 77, 334–338.

JEDEC (2011). JEDEC JEP-122G Failure Mechanisms and Models for Semiconductor Devices. JEDEC Solid State Technology Association, Arlington, VA.

Jin, D. and Del Alamo, J. (2012). Impact of high-power stress on dynamic ON-resistance of high-volatge GaN HEMTs. *Microelectronics Reliability Journal*, 52, 2875–2879.

Lall, P., Pecht, M., Hakim, E.B. (1997). *Influence of Temperature on Microelectronics and System Reliability*. CRC Press, Boca Raton, FL.

Liu, C., Sagong, H.-S., Kim, H., Choo, S., Lee, H., Kim, Y. (2015). *Systematical Study of 14 nm FinFET Reliability: From Device Level Stress to Product HTOL*. IRPS, Monterey, CA.

Loubet, N., Hook, T., Montanini, P., Yeung, C., Kanakasabapathy, S., Guillom, M., Yamashita, T., Zhang, J., Miao, X.H., Wang J. et al. (2017). Stacked nanosheet gate-all-around transistor to enable scaling beyond FinFET – T17-5. *Symposium on VLSI Technology Digest of Technical Papers*, Kyoto, Japan, June.

McPherson, J.W. (2019). *Reliability Physics and Engineering: Time-to-Failure Modeling*, 3rd edition. Springer Nature Switzerland AG, Plano, TX.

McPherson, J.W. and Baglee, D. (1985). Acceleration factors for thin gate oxide stressing. *Proc. 23rd Annual IEEE IRPS*, p. 1.

Meneghesso, G., Meneghini, M., Tazzoli, A., Ronchi, N., Stocco, A., Chini, A., Zanoni, E. (2010). Reliability issues of gallium nitride HEMTs. *International Journal of Microwave and Wireless Technologies*, 2(1), 39–50.

Meneghini, M., Meneghesso, G., Zanoni, E. (2017). Power GaN devices: Materials, applications and reliability. In *Power Electronics and Power Systems*, Chow, J.H., Stankovic, A.M., Hill, D. (eds). Springer, Berlin.

National Academy of Sciences (2008). Panel on reliability growth methods for defense; Reliability growth: Enhancing defense system reliability. *The National Academies Press*, Washington, DC.

Nikulin, M., Couallier, V., Gerville-Reache, L. (2009). Accelerated life and degradation models in reliability and safety: An engineering perspective. *EA 2961, Statistique Mathématique et ses Applications*, Univ. Bordeaux 2, Bordeaux.

PRISM Now 217Plus (TM) (n.d.). RIAC – The Reliability Analysis Center [Online]. Available at: http://www.theriac.org.

QUANTERION (2015). *Handbook of 217Plus Reliability Prediction Models*. QUANTERION Solution Inc, Utica, NY.

Robustness Group ZVEI (2013). *Handbook for Robustness Validation of Automotive Electrical/Electronic Modules*. ZVEI, Frankfurt am Main.

Sedyakin, N. (1966). On one physical principle in reliability theory. *Techn. Cybernetics*, 80–87.

Serma Technologies (1996). Alcatel Espace Internal failure analysis report EL 4103. Report, Serma Technologies, Toulouse, March.

Snow, E., Grove, A., Deal, B., Sah, C. (1965). Ion transport phenomena in insulating films. *Journal of Applied Physics*, 36, 1664.

Srinivasan, J. and Adce, S.V. (2003). The case for microarchitectural awareness of lfetime reliability. Technical report, UIUC CS n° UIUCDCS-R-2003-2391.

Stathis, J., Mahapatra, S., Grasser, T. (2018). Controversial issues in negative bias temperature instability. *Microelectronics Reliability Journal*, 81, 244–251.

Suhir, E. (2010). Probabilistic design for reliability. *ChipScale Reviews*, 14(6).

Suhir, E. (2013). Predicted reliability of aerospace electronics: Application of two advanced concepts. *IEEE Aerospace Conference*, 2–9 March.

Takeda, E. and Suzuki, N. (1983). An empirical model for device degradation due to hot-carrier injection. *IEEE Electron Device Letters*, EDL-4, 111–113.

Verkest, D. (2018). /cadence_blogs_8/b/breakfast-bytes/posts/imec2. IMEC Roadmap – Cadence Breakfast Bytes Blogs – Post from Paul McLellan, 7 June [Online]. Available at: https://community.cadence.com/ [Accessed 28 February 2019].

White, M. and Bernstein, J.B. (2008). Microelectronics reliability: Physics-of-failure based modeling and lifetime evaluation. JPL Publication 08-5, Jet Propulsion Laboratory/ California Institute of Technology.

Wigner, E. (1938). The transition state method. *Transactions of the Faraday Society (London)*, 34, 29–41.

1

Basic Reliability Tools for SHM Protocols

As far as the laws of mathematics refer to reality, they are not certain; and as far as they are certain, they do not refer to reality.

Albert EINSTEIN
Address to the Prussian Academy of Sciences, January 27, 1921

1.1. Introduction

Trending articles and news related to the semiconductor industry have been published on various websites. Khaveen Jeyaratnam, a specialist in financial analysis, valuation and investment research, presented the semiconductor industry value chain and discussed the top three semiconductor foundries in the world in a paper presented in March 21, 2019[1]. He highlighted that:

> [...] the semiconductor industry is experiencing heavy growth in demand from the onset of the fourth wave of technology which includes robotics, artificial intelligence, nanotechnology, quantum computing, biotechnology, the Internet of Things, fifth-generation wireless technologies (5G), augmented/virtual reality, 3D printing, and autonomous vehicles.

The top four worldwide semiconductor foundries by revenue for 1H18 are TSMC (56.1% market share), GlobalFoundries (9.0%), UMC (8.9%) and Samsung (7.4%). Currently, TSMC ($32B+) is the number one foundry by a very large margin, with GF ($6B+), UMC (>$5B+), Samsung (>$5B+) and SMIC ($3B+)

Chapter written by Alain BENSOUSSAN, Joseph B. BERNSTEIN and Alain BRAVAIX.

1 See: https://seekingalpha.com/article/4250209-top-3-semiconductor-foundry-stocks-buy.

barely visible in the rearview mirror. The Semiwiki website presents the full year 2018 revenues of TSMC broken down into the various technologies, with 9% for the 7-nanometer node size (25% in 2019), 11% for the 10-nanometer, 23% for the 16/20-nanometer and 63% for the 28-nanometer and below. On the other side, GlobalFoundries offer CMOS, FinFET and FD-SOI technologies.

Electronic components are made for devices, including equipment to complete systems, and are fabricated from materials and structures that degrade over time under normal operational conditions. This is a fundamental consequence of the second law of thermodynamics which cannot be avoided in our physical world. We cannot afford uncertainty of systems within various domains including aeronautics, more electric aircraft (MEA), space, automotive and autonomous vehicles, all of which rely on hundreds, thousands or even millions of electronic devices per vehicle and and which employ fleets of thousands. Degradation has also been observed to depend on many environmental parameters, both intrinsic and extrinsic, due to:

– packaging and component manufacturing and process distributions;

– original material and geometry arrangement in devices;

– internal and external environmental conditions, referred to as "*stressors*", such as biasing (voltage, current, power dissipation, transient, signal level), temperature (low, high, constant, variable), irradiation levels (particles, X-ray, γ-ray), humidity, other pollution gases, etc. (JEDEC 2011; McPherson 2019a).

We can model failure rates and degradation, assuming independently thermal activation processes as well as at random, until a particular critical device parameter can no longer meet the required specification for proper device or system functionality. Some parameters (considered as electrical signatures) can be identified as a **precursor** of failure occurrence. So, the failure is really defined when a specific functional parameter is affected "enough" to create a malfunction in the system. The threshold, for an indicator degradation leading to a final system parametric failure or a catastrophic failure, depends on the stressor intensities (level of stresses with respect to the intrinsic limits supported by a device) and time.

Much statistics and mathematics has been developed over the centuries (Boltzmann 1886; Evans and Polanyi 1938; Black 1969; Eyring et al. 1980; Berthon et al. 2014; McPherson 2019a) to describe failure rates and the well-known bathtub curve showing a schematic of failure rate behavior over time. In this approach, three domains were traditionally established: infant mortality, random failure and wearout failure (SIEMENS AG 1986; RDF 2000; National Academy of Sciences 2008). For example, there is the common "time-to-failure" (TTF) parameter, the time for a given population to survive in order to have its number of "good" elements be half the original population (McPherson 2019a).

This simplified viewpoint, which has drawbacks and suffers from some vagueness in the definition of failure when considering innovative commercial-off-the-shelf (COTS) devices use (for deep submicron technology with node size lower than 20 nm (Huard et al. 2012) or wide band-gap semiconductor-based devices (Quay 2008)) the following:

– Infant mortality, random failures and wearout are generally based on catastrophic failure observations in the field. When considering parametric failures, this may induce different interpretations and results (Bernstein and Bensoussan; INTEL 2002; Bernstein et al. 2014).

– Temperature definition is a key parameter which can modify the interpretation and the reading of results (Lee et al. 1981; Lall et al. 1997): for example, temperature can be considered as ambient or case or junction.

– Failure criterion definition is observed to be a sensitive parameter impacting the TTF (Bensoussan et al. 1994; Decker 1994; Delaney et al. 1994).

Saying this, the broad application of disruptive technologies in embedded Hi-Rel systems requires all reliability aspects to be modeled and quantified in order to predict TTF and wearout remaining useful life (RUL), both impacted by the mission profile of electronic systems.

This chapter will first detail the existing models, tools and methodologies existing for COTS new emerging technologies. We will see why and how the concept of activation energy is an Eyring model related to the stress and temperature applied, and the use of an approximated constant activation energy can no longer be considered as a constant to extrapolate some experiments under high stress to a nominal mission operation profile.

We present hypotheses, baselines and definitions of reliability parameters through the prism of transition state theory (TST) in order to show how the multi-dimensional stress environment can be simply modeled. Reliability prediction is then supported by the extrapolation of lifetime models based on what we call Physics-of-Healthy (PoH). PoH, in terms of reliability prediction, is the calculation of the statistical failure probability described by physics-based rate processes. The methodology presented here is a direct application of the well-known, standard, proportional sum-of-failure-rates where the rates are calculated by thermodynamically determined statistical processes.

One section will address the transition state reliability model and reliability PoH defined as the probability of a product performing its intended function for a given amount of time and mission profile. Basic approaches are recalled so that engineering designs and materials are detailed to construct reliability tools, as defined by conventional reliability standards.

The final section will focus on system reliability (series and parallel) mathematical reminders and proposed approximations through four Lemmas.

L1: a series system constituted of n identical and independent elements, each described by a POISSON distribution (λ_0) reliability model that can be approximated by a general equivalent POISSON distribution with parameter $n \cdot \lambda_0$.

L2: a parallel system constituted of n identical and independent elements, each described by a POISSON distribution (λ) reliability model that can be approximated by a general equivalent WEIBULL distribution with parameters λ and $\beta = n$ with an error lower than 1% for time operation lower than 30%·MTTF for the example shown.

L3: a parallel system constituted of n identical and independent elements, each described by a reliability WEIBULL distribution (α, β) model that can be approximated by a general equivalent WEIBULL distribution with parameter ($\alpha, \beta \cdot n$) with an error lower than 1% for time operation lower than 75%·MTTF.

L4: in a series-parallel system constituted of n identical and independent elements, each described by a reliability POISSON distribution (λ), its reliability model can be approximated by a general equivalent WEIBULL distribution with parameter ($\theta, \beta = m$) with an error lower than 3% for time operation lower than 50%·MTTF.

The fact that microelectronic devices are manufactured in such large quantities, with established variability, deviations (fluctuations) and uncertainties, makes them the quintessential framework through which all reliability prediction and evaluation should be understood. Existing reliability models may sometimes be less accurate for emerging and non-mature technologies because of a lack of customer and user feedback due to the short product lifecycle. The consumer market relies on short lifecycles. Consequently, the very nature of the electronics industry, where there are so many parts having so many interacting physical phenomena, electrical performances, geometrical and material parameters, and application design conditions, makes them the ideal example for reliability evaluation that can be applied to any other product or industry.

Part 1 of this book prepares the foundation to answer three central questions:

1) How do we apply reliability prediction tools to innovative technologies?

2) How are the hypotheses constructed and how is the methodology implemented in these tools, and to what extent are they able to cover emerging technologies and applications?

3) Prognostic failure model (PFM): what could be a proposed guideline to organize all the concepts described?

1.2. State-of-the-art reliability in DSM and GaN technologies and Physics-of-Healthy: thermodynamics

1.2.1. *COTS and emerging technologies in deep-sub-micron technologies: short overview*

In a recent paper, Jones (2019) presented highlights on NAND scaling with layers, DRAM peripheral scaling and logic high performance and IoT. Evidence is shown on the complexity of the market push technology in the 3D NAND mask counts and bit density trends.

> The transition from 2D NAND to 3D is enabling the continuation in bit density scaling by using the third dimension [see Figure 1.1]. DRAM scaling is limited by capacitor size and facing physical limits. Logic is continuing to scale but fundamental limits on 2D shrinks are looming. Leading edge logic has evolved from planar transistors to a split roadmap with FinFET for high performance and things like FDSOI for IOT. Longer term gate-all-around is on the horizon.

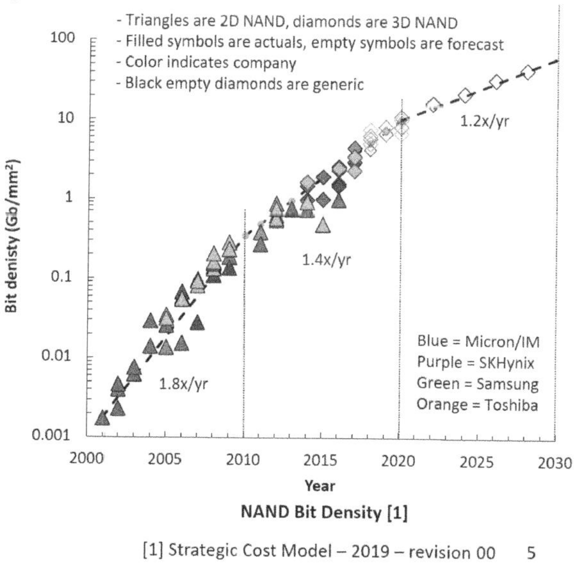

Figure 1.1. *NAND bit density (courtesy of IC Knowledge LCC in SemiWiki (Jones 2019)). For a color version of this figure, see www.iste.co.uk/elhami/mechatronics.zip*

From Figure 1.1, we can see that for planar transistors, the effective gate width (Weff) was defined by the transistor gate width. For FINFETs, the Weff depends on the width and 2x the height of the fin and therefore the Weff can only be varied in discrete increments. With GAA, the Weff is 2x thickness and 2x width.

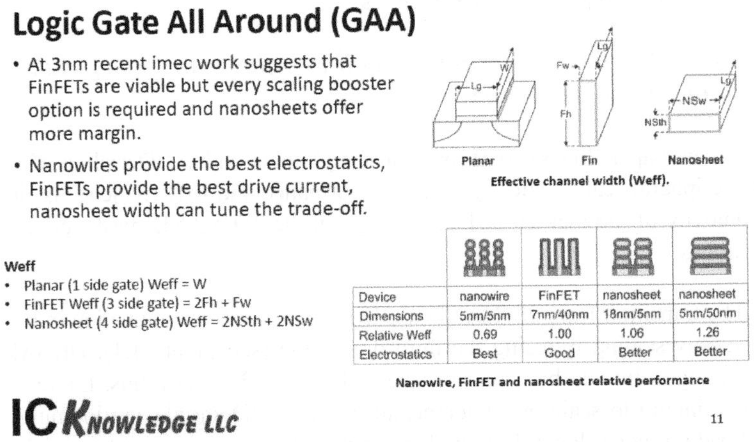

Device	nanowire	FinFET	nanosheet	nanosheet
Dimensions	5nm/5nm	7nm/40nm	18nm/5nm	5nm/50nm
Relative Weff	0.69	1.00	1.06	1.26
Electrostatics	Best	Good	Better	Better

Figure 1.2. *Logic gate all around (GAA) size definition according to ref (Jones 2019). For a color version of this figure, see www.iste.co.uk/elhami/mechatronics.zip*

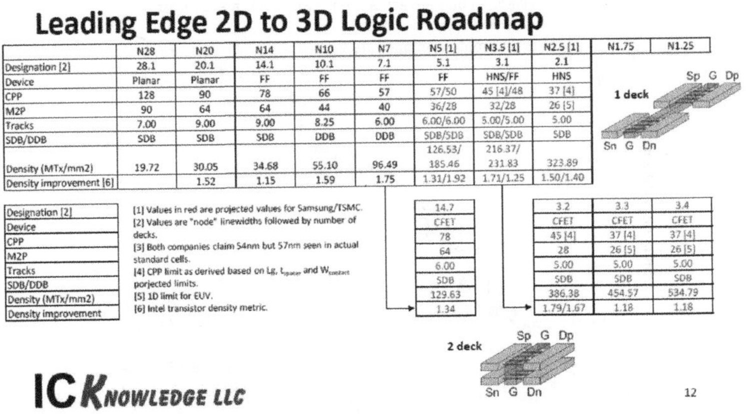

	N28	N20	N14	N10	N7	N5 [1]	N3.5 [1]	N2.5 [1]	N1.75	N1.25
Designation [2]	28.1	20.1	14.1	10.1	7.1	5.1	3.1	2.1		
Device	Planar	Planar	FF	FF	FF	FF	HNS/FF	HNS		
CPP	128	90	78	66	57	57/50	45 [4]/48	37 [4]		
M2P	90	64	64	44	40	36/28	32/28	26 [5]		
Tracks	7.00	9.00	9.00	8.25	6.00	6.00/6.00	5.00/5.00	5.00		
SDB/DDB	SDB	SDB	SDB	DDB	DDB	SDB/SDB	SDB/SDB	SDB		
Density (MTx/mm2)	19.72	30.05	34.68	55.10	96.49	126.53/ 185.46	216.37/ 231.83	323.89		
Density improvement [6]		1.52	1.15	1.59	1.75	1.31/1.92	1.71/1.25	1.50/1.40		

Designation [2]	14.7	3.2	3.3	3.4
Device	CFET	CFET	CFET	CFET
CPP	78	45 [4]	37 [4]	37 [4]
M2P	64	28	26 [5]	26 [5]
Tracks	6.00	5.00	5.00	5.00
SDB/DDB	SDB	SDB	SDB	SDB
Density (MTx/mm2)	129.63	386.38	454.57	534.79
Density improvement	1.34	1.79/1.67	1.18	1.18

Figure 1.3. *A roadmap from 2D planar transistors at the 28 nm and 20 nm nodes to FinFETs and then HNS and eventually stacked 3D CFETs (Jones 2019). For a color version of this figure, see www.iste.co.uk/elhami/mechatronics.zip*

From Figures 1.2 and 1.3 for node 5 (N5) and N3.5, we have specific projections for Samsung and TSMC. At N2.5, we have a generic forecast with both companies converged on HNS.

For 3D, we have options, starting with a relaxed 14 nm design rule CFET with seven layers as well as more lithographically aggressive 3.2–3.4 complementary transistors (CFETs) with 3 nm lithography and 2, 3 and 4 layers. The figures show an nFET and pFET side by side for a single-layer device and then a pFET over and nFET for a two-layer CFET.

According to Jones, it is shown how the EUV helps to mitigate mask count increases and about 60–70 mask sets. CFET at a 1.75 nm node also helps to control the lithographic difficulty by being highly self-aligned.

Traditional device scaling continues, but dimensional scaling also impacts performance, as shown in Figure 1.4, an extract from the Cadence blog Paul McLellan published in June 2018.

> Fundamental new devices such as carbon nanotubes or 3D materials are still far from maturity. So, there have been three eras. In the first, the focus was to scale the devices and wires. In the second, the focus was reducing the size of basic logic cells (and the SRAM cell). In the future, the focus has to move to scaling entire sub-system functions.

According to Dan Mocuta (Director of logic integration and devices at Imec), quoted in (Lapedus 2019): "the key element is variable widths. You can control it better than the variable height of a fin". He further commented:

> In a FinFET technology, the width of the device is quantized. You can have one fin, two fins, three fins or whatever. In nanosheets you have a fixed number of nanosheets on top of each other. But you can play with the width. Now, you have access to a continuum of device widths, which you didn't have for the FinFET [...] For example, you want to have an area that drives a lot of current. That could be a buffer. Then, you want to have an SRAM with a very small footprint. There are different needs on the chip that can be met.

Nanosheets are promising, but this is not the only option (Figure 1.5). With a breakthrough, finFETs could extend beyond 5 nm. Another option is to wait until the industry develops a better transistor. Yet another way is to get the benefits of scaling by putting multiple devices in an advanced package.

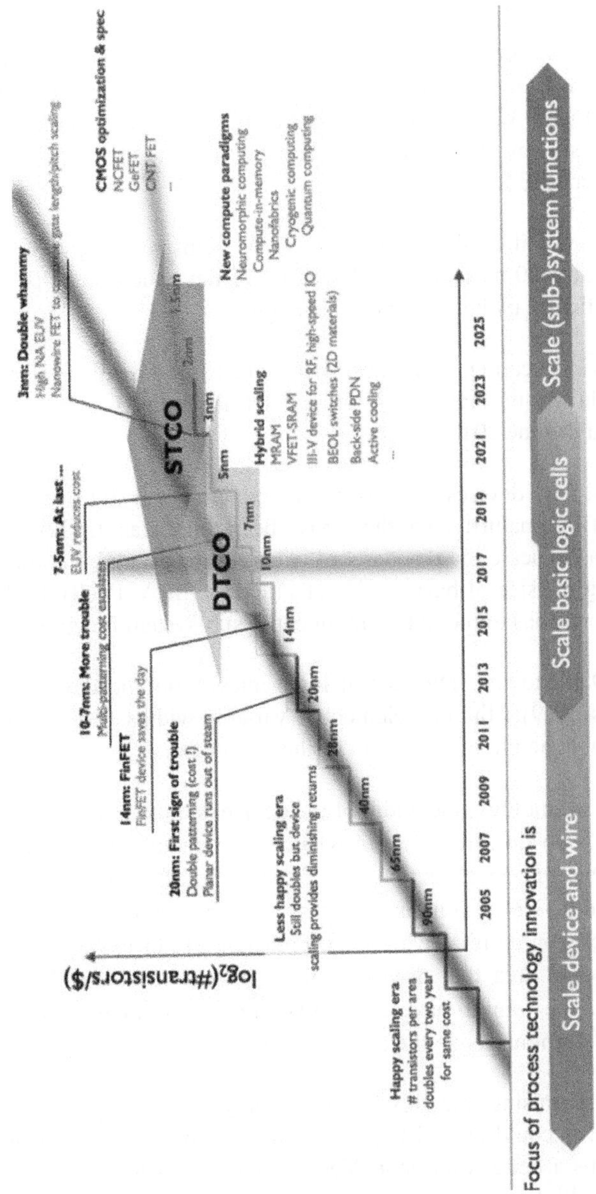

Figure 1.4. *A roadmap from IMEC by Verkest (2018). For a color version of this figure, see www.iste.co.uk/elhami/mechatronics.zip*

For now, gate-all-around technology appears to be the most practical technology after finFETs.

Modeling such novel transistors to establish reliability figures for risk assessment is tremendously challenging. The market will be driven by system application and deep submicron technology as a "COTS integration plug and play" will require risk assessment methodology to be implemented to satisfy markets related to automotive, aeronautic and space where system failures in normal operation and harsh mission profile conditions are strictly forbidden for a certain period of time. This has clearly not addressed the next question related to MTTF or MTBF for system repairs.

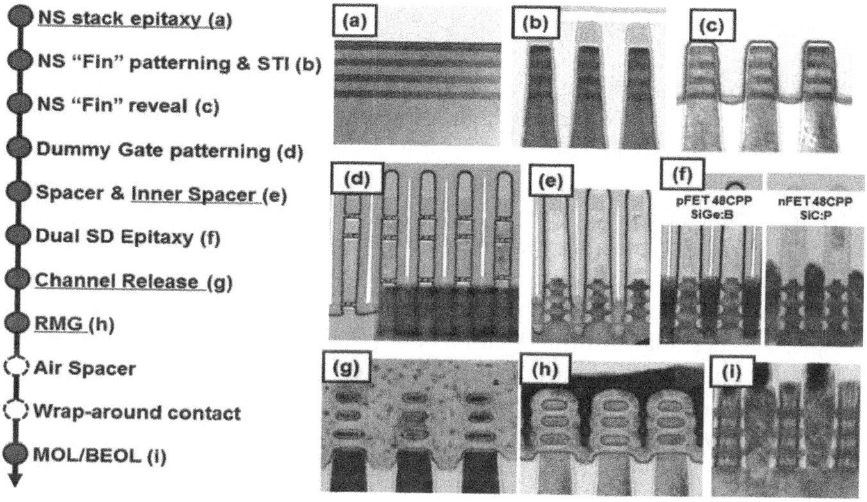

Figure 1.5. *Stacked nanosheet process sequence and TEM. IBM, Samsung Electronics, Global Foundries (Loubet et al. 2017)*

Having established that the hardware is particularly challenging, we know any system and component in a custom or niche application will tend to degrade over time when operating under normal or even sometimes extreme conditions. However, the most reliable technology will continue to be affected by changes in their intrinsic characteristics and performance or due to the extrinsic high stress applied.

1.3. General overview on GaN device failure mechanisms

Normally-off high-power switching AlGaN/GaN-on-Si heterojunction transistors are promising for high reliability space applications. Nevertheless, such and similar

GaN technologies suffer from well-known failure mode issues as studied since the last decade by several authors on gate current leakage increase and/or current collapse (Ueda 2014; Meneghini et al. 2016), dynamic on-resistance with recovery effects (Jin and Del Alamo 2012), strain relaxations and trap charging effects (Meneghesso et al. 2008; Wu et al. 2014), TDDB mainly related to GaN MIS-HEMTS structures (Wamock and Del Alamo 2017) and PBTI of GaN MOSFETs (Guo and Del Alamo 2015). In Meneghini et al. (2017) (Chapter 13), Joachim Wuerfl (FBH, Berlin) describes the most relevant drift effects that limit the performance of GaN-based power transistors. Degradation and drift effects on semiconductor devices can affect their performance either reversibly or definitively. Many failure mechanisms have been identified and described, which demonstrates that they are imperative for predicting device performance in a real-system environment.

The effect of temperature on electronic devices is often assessed by extrapolating from accelerated tests at extremely high temperatures based on the Arrhenius law. This method is known to be not necessarily accurate for prediction, particularly when stress-induced failures are driven by non-thermal dynamic electrical stresses. Theoretical work in kinetic theory, thermodynamics and statistical mechanics has developed forms that contain exponential forms similar to Arrhenius. It is observed that the well-known Arrhenius law usually does apply, albeit with some modification, within existing models describing physics of failure. This is known, by example, in Black's law, Coffin–Manson or any application of Eyring's law. These include the effect of humidity or the hydrogen poisoning or other effects in semiconductors (Black 1969; Agarwala et al. 1970; Eyring et al. 1980; Decker 1994; Kayali et al. 1996; JEDEC 2011, 2017). When these effects are simultaneously activated under multiple stress conditions inducing a series of different failure modes and mechanisms, the standard reliability predictive models are questionable.

Hence, regarding the normally-off eGaN-on-Si heterojunction transistors, a critical concern with such new technology is reliability. It is mandatory to study the role of temperature and biasing stress on the high-power degradation of this type of technology.

A number of drift effects are influencing the performance of power switching devices, reducing switching efficiency and potentially compromising reliability. Figure 1.6 as well as Figures 1.7 and 1.8 summarize the most important reversible drift mechanisms observed in GaN transistors. Often, the effects described above are correlated with each other resulting in a great variety of device parameter changes.

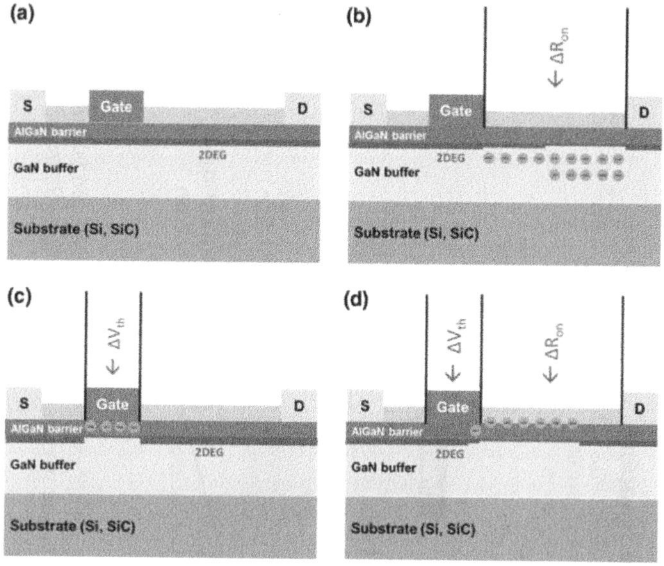

Figure 1.6. *Schematic cross-section of an AlGaN/GaN HEMT at different bias and trapping conditions (from (Meneghini et al. 2017)). For a color version of this figure, see www.iste.co.uk/elhami/mechatronics.zip*

a) Ideal device without any trapping at on-state: the 2DEG is fully populated with electrons.

b) Electron trapping in the vicinity of the drain access region: this situation, for example, happens immediately after switching from off-state to on-state at high drain bias. Trapped electrons or depleted deep acceptors completely or partially deplete the 2DEG and hence impede current flow. If the source or the drain access regions are influenced by trapping, the on-state resistance changes. This gives rise to the so-called dynamic on-state resistance increase ΔR_{on_dyn}.

c) Electron trapping in the vicinity of the gate (underneath the gate) results in a partial depletion of the channel just underneath the gate: as these charges aid depleting the channel, a lower gate voltage is necessary to fully turn off the device – the threshold voltage shifts to more positive values.

d) Electron trapping at the interface to passivation layer and in the barrier layer: this leads to both a threshold voltage shift and to the formation of a virtual gate at the interface oriented towards the drain reduction of 2DEG electron concentration, R_{on_dyn} increase.

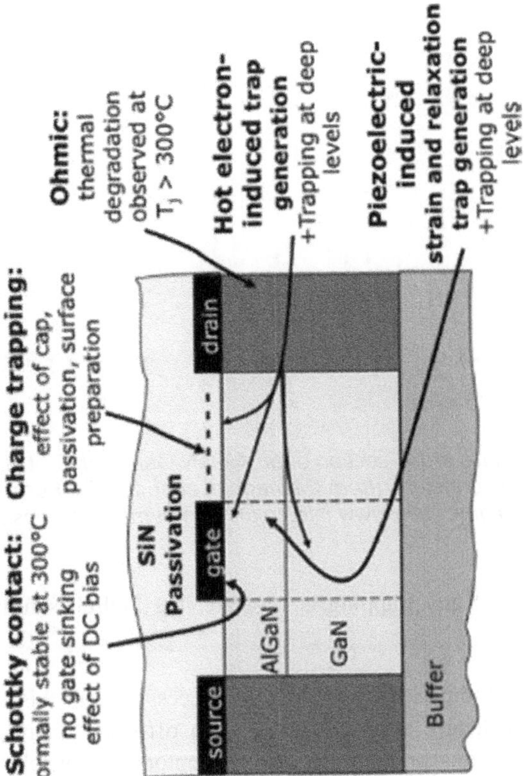

Figure 1.7. *Schematic cross-section of an AlGaN/GaN HEMT RF transistor, identifying critical areas which can be subjected to degradation (Meneghesso et al. 2008). For a color version of this figure, see www.iste.co.uk/elhami/mechatronics.zip*

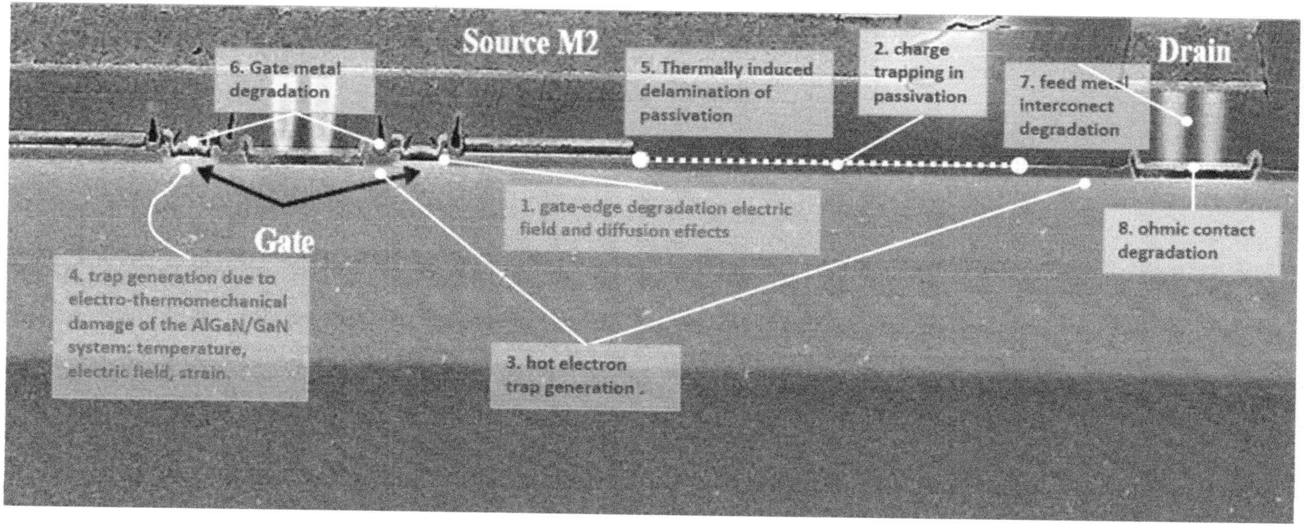

Figure 1.8. *Failure mechanisms recently identified on GaN HEMTs. Mechanisms identified in red (5–8) are thermally activated mechanisms. Mechanisms 2 and 3 are related to the presence of hot electrons present at high bias conditions. Mechanisms 1 and 4 (green) are particular to GaN devices due to the polar and piezoelectric nature of this semiconductor (freely arranged from Meneghesso et al. 2010). For a color version of this figure, see www.iste.co.uk/elhami/mechatronics.zip*

Negative charges trapped in the source or drain access region will reduce the 2DEG electron density, resulting in an increase in on-state resistance. For traps confined to regions underneath the gate and in the presence of negative charges the equilibrium must be compensated by a positive threshold voltage shift; collaterally, threshold voltage can also shift in the negative direction if positive charges are introduced (e.g. if electrons are emitted from donor traps).

For GaN RF devices, Meneghesso degradation (Meneghesso et al. 2008) can still be observed on chosen parameters named "*indicators*" identified as key parameters showing pre-evolution of the future performance of a device.

1.4. Physical reliability models applied to DSM technology

In a recent paper, Yunm et al. (2018) suggest several entropies as candidates for proper damage measurements in the fatigue process. The entropic approaches are expected to develop in order to apply them in practical reliability analyses and prognoses. The paradigm of the TST developed by Wigner (1938) and by Evans and Polanyi (1938) is an approach we can benefit from if we adapt to the concept of a unified semiconductor reliability model. Indeed, in the early 20th century, the TST was applied to chemistry transformations by Eyring et al. (1980) and Glasstone et al. (1941). The TST was developed in chemistry based on Hammond's postulate, which was published in 1955 and applied to physical organic chemistry (Hammond 1955).

From the 1960s and over the 40 years that followed, several authors reported on complementary reliability models centered around the mobile-ion (Na+) drift within dielectric (Snow et al. 1965), to intrinsic mechanisms such as TDDB (McPherson et al. 2000a); for example, while in the mid-1980s, McPherson and Baglee (1985; McPherson 1986) described stress-dependent activation energy and developed a generalized Eyring model in order to better understand thermally activated failure mechanisms (McPherson 2019a).

Similarly in 2013, the stress-dependent activation energy model based on the Boltzmann–Arrhenius–Zhurkov (BAZ) model for multiple stress proposed by Suhir (Suhir 2010, 2013; Bensoussan and Suhir 2013) presented a generalization of the principle proposed by Zhurkov (1965) as well as Cox (1972; Cox and Oakes 1984). Suhir suggests considering any type of stress (not only mechanical but also electrical or external) as key parameters to modify the equivalent activation energy of a given failure mechanism. In other words, the activation cannot be considered as a constant parameter with respect to any type of stress (temperature range, biasing and signal stress range). We can view this approach as an equivalent TST concept applied to the reliability paradigm (Suhir 2013).

Multiple failure mechanisms and physics of degradation in semiconductors may occur in a single set of TTF data but without obvious points of inflection to help separate the mechanisms. McPherson, in the third edition of his book *Reliability Physics and Engineering*, provides the basics of reliability modeling (McPherson and Baglee 1985; McPherson 2019a):

> [...] generally, materials/devices exist in metastable states. These states are referred to as being metastable because they are only apparently stable. Metastable states will change/degrade with time. The rate of degradation of the materials (and eventual TTF for the device) can be accelerated by an elevated stress (e.g. mechanical stress, electrical stress, electrochemical stress, etc.) and/or elevated temperature.

The Gibbs free energy description of material/device degradation is illustrated in Figures 1.13 and 1.14. Let us consider the initial state to be a sound device before aging and the final state of a degraded device (either catastrophically failed or degraded and not compliant with the acceptable performance limit of the device). It is important to note at this point that the reliability model is described by parameter drift degradation as a function of time, and not as a random failure paradigm. In this diagram, the net reaction rate is a dynamic equilibrium between forward and reverse reactions meaning the degradation could be reversible. The net reaction process is written as:

$$k_{Net} = k_{Forward} - k_{Reverse} \qquad [1.1]$$

with $k_{forward}$ and $k_{reverse}$ as functions of temperature and stress S applied:

$$k_{Forward} = exp\left(-\frac{\Delta G_F^{\ddagger}(S,T)}{k \cdot T}\right) \qquad [1.2]$$

and

$$k_{Reverse} = exp\left(-\frac{\Delta H_R^{\ddagger}(S,T)}{k \cdot T}\right) \qquad [1.3]$$

where T is the temperature in Kelvin, k is the Boltzmann constant, and $\Delta G^{\ddagger}{}_F(S, T)$ represents the free energy of activation associated with the reaction process for the forward reaction. $\Delta H^{\circ}{}_R(S, T)$ represents the change in enthalpy required for the final state reached for the reverse reaction. ΔS° is the entropy change, for example, the driving force for device degradation.

Figure 1.13 represents the simplified free Gibbs energy diagram for a reaction activated only by the temperature for a sound device to a degraded device and vice versa. Let us assume a device placed in an electronic system designed to implement an electronic function for a given application. The electronic function is supposed to fulfil a mission for a given life mission without failure, meaning without exceeding performance limits defined by the system functionality. In such conditions, a degraded device is defined as a device whose characteristics have changed to a certain extent above a given limit criterion fixed for the system to function properly. Let us call the final state the state where the device reaches the failure criteria. So, equations [1.1], [1.2] and [1.3] describe the probability the transition rates are associated with a "sound device" to reach a failure limit.

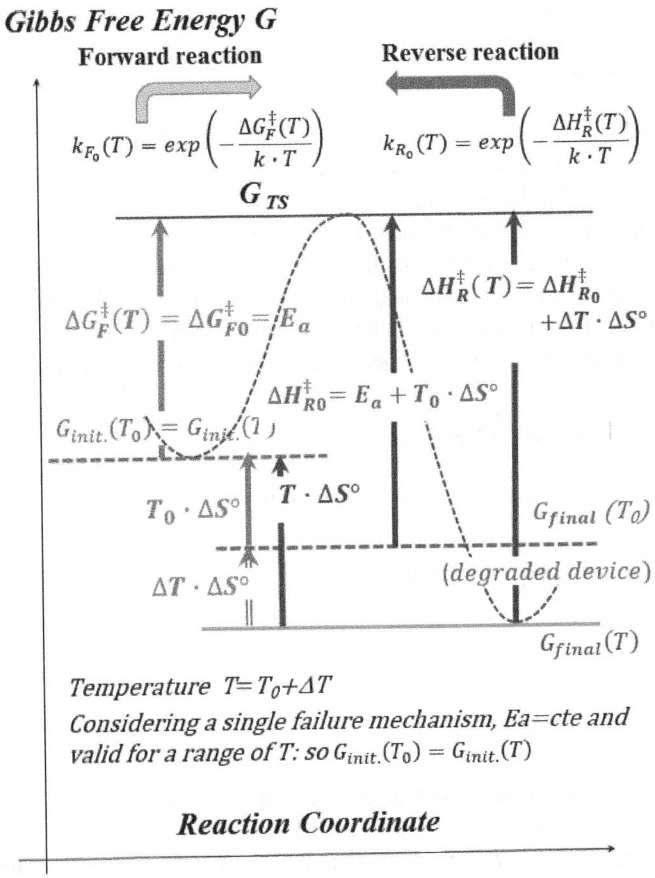

Figure 1.9. *Gibbs free energy diagram for T temperature increase applied, for example, Arrhenius thermal activation energy diagram. For a color version of this figure, see www.iste.co.uk/elhami/mechatronics.zip*

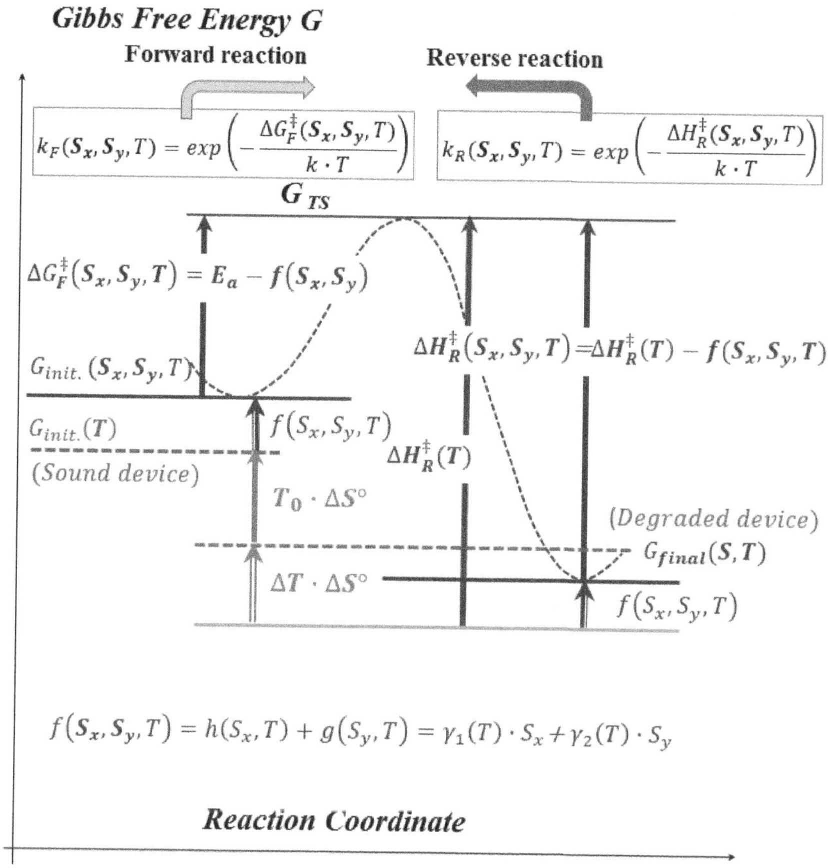

Figure 1.10. *Gibbs free energy diagram considering reaction with two external stresses Sx and Sy applied and temperature T (diagram modified from (McPherson 2019a)). For a color version of this figure, see www.iste.co.uk/elhami/mechatronics.zip*

Figure 1.10 shows the Gibbs free energy diagram when considering multiple stress conditions, depicting how they impact the diagram. Note that the equivalent activation energy $\Delta G^{\ddagger}_F(S, T)$ is reduced when stresses are applied and can reduce to zero for an extreme stress limit. So, a stress current may conduct to an instantaneous burnout catastrophic failure. As a result, the energy barrier vanishes because of a catalyst effect and the stress function $f(I_{burnout})$ compensates exactly the initial $\Delta G^{\ddagger}_{F0} = E_a$ at a given temperature T. So, the equivalent activation energy ranges from E_a (*the Arrhenius pure thermal value*) to 0 for a current stress range for $0 \leq I \leq I_{burnout}$ or for $0 \leq I/Iburnout \leq 100\%$. Similarly, we can assume multiple stress to cumulate and the combination of them may reach the instantaneous catastrophic failure when the initial Arrhenius activation energy is fully compensated. These conditions can be compared when considering current, voltage and power dissipation parameters of a transistor defined by their limits called the safe operating area, as shown in the example given in Figure 1.11. The static characteristic limits related to current (vertical axis), voltage (horizontal axis) and DC or pulsed power dissipation are plotted in the log–log scale. In this case, the *I(V)* curves are limited by the power dissipation capability of the device (straight lines at the top right corner of the figure). The burnout can be reached either under high current or high voltage or high power overstresses. The combination of current and voltage stresses is seen through the power dissipation parameter: either one or the other or both lead to the destruction of the device. This paradigm can also be modeled by Figure 1.10. It is worth considering these parameter limits as static limits but similarly, dynamic and pulsed operations can be set.

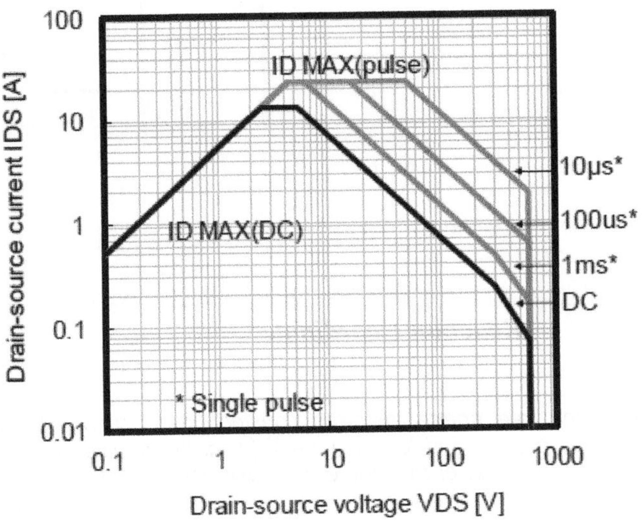

Figure 1.11. *Example of maximum rating and safe operating area plot for a power transistor*

We can observe the enthalpy $\Delta H_R^{\ddagger}(S=0,T_0)$ is by definition independent of the temperature variation ΔT because the transition state G_{TS} as well as the final state G_{final} is only related to the initial T_0 value. We assume that the expression of the enthalpy is:

$$\Delta H_R^{\ddagger}(S_i=0,T_0)=\Delta H_{R0}^{\ddagger}=E_a+T_0\cdot\Delta S^{\circ}\ \text{for i = 1 , n} \quad [1.4]$$

Let us consider the case to superpose n stress conditions S_i. For the simplest case of $n = 2$ stressors S_x and S_y, a generalized Eyring model was established in (McPherson 1986), describing thermally activated failure mechanisms in materials/devices under stress. The forward reaction described by the Gibbs free energy (respectively enthalpy for the reverse reaction) is a function of the generalized stress S, and we set a Taylor series expansion around point $(S_x, S_y) = (S_a, S_b)$ expanded to (limited to the first order):

$$f(S_x, S_y, z) = f(S_a, S_b, c) + \frac{\partial f(S_a, S_b, c)}{\partial S_x} \cdot (S_x - S_a) + \frac{\partial f(S_a, S_b, c)}{\partial S_y} \cdot (S_y - S_b) \qquad [1.5]$$

We consider the two infinitely differentiable functions $\Delta G^{\ddagger}_F\ (S_x, S_y, T)$ and $\Delta H^{\ddagger}_R\ (S_x, S_y, T)$, limited to the Taylor series expansion linear around $(S_x, S_y) = (S_{a0}, S_{b0})$:

$$\Delta G^{\ddagger}_F(S_x, S_y, T) \approx \Delta G^{\ddagger}_F(S_{a0}, S_{b0}, T_0) + \left[\frac{\partial\left(\Delta G^{\ddagger}_F(S_x, S_y, T)\right)}{\partial S_x}\right]_{S_{a0}, S_{b0}} \cdot (S_x - S_{a0}) + \left[\frac{\partial\left(\Delta G^{\ddagger}_F(S_x, S_y, T)\right)}{\partial S_y}\right]_{S_{a0}, S_{b0}} \cdot (S_y - S_{b0}) \qquad [1.6]$$

$$\Delta H^{\dagger}_R(S_x, S_y, T) \approx \Delta H^{\dagger}_R(S_{a0}, S_{b0}, T_0) + \left[\frac{\partial\left(\Delta H^{\ddagger}_R(S_x, S_y, T)\right)}{\partial S_x}\right]_{S_{a0}, S_{b0}} \cdot (S_x - S_{a0}) + \left[\frac{\partial\left(\Delta H^{\ddagger}_R(S_x, S_y, T)\right)}{\partial S_y}\right]_{S_{a0}, S_{b0}} \cdot (S_y - S_{b0}) \qquad [1.7]$$

According to Figures 1.9 and 1.10, we observe:

$$\Delta G^{\dagger}_{F0}(T_0) = E_a \quad \text{and} \quad E_a = \Delta H^{\ddagger}_{R0} - T_0 \cdot \Delta S^{\circ} \qquad [1.8]$$

$$\Delta H^{\ddagger}_R(S_x, S_y, T) = E_a + T \cdot \Delta S^{\circ} - \frac{1}{2} \cdot f(S_x, S_y, T) \qquad [1.9]$$

$$\Delta G^{\ddagger}_F(S_x, S_y, T) = E_a - \frac{1}{2} \cdot f(S_x, S_y, T) \qquad [1.10]$$

Let us assume simply

$$f(S_x, S_y, T) = h(S_x, T) + g(S_y, T) \qquad [1.11]$$

The factor ½ is set because we consider the same stress level affecting the entropy on the initial state and the enthalpy affecting the final state.

At zero stress, the activation energy is converging to the standard Arrhenius thermal activation energy, so we must assume $f(S_x=0, S_y=0) = 0$ and then

$$\boldsymbol{h(S_x = 0, T) = 0} \text{ and } \boldsymbol{g(S_y = 0, T) = 0} \tag{1.12}$$

According to the Eyring law, the stress function $f(S_x, S_y)$ is defined by:

$$\boldsymbol{h(S_x, T) = \gamma_1(T) \cdot S_x} \tag{1.13}$$

and

$$\boldsymbol{g(S_y, T) = \gamma_2(T) \cdot S_y} \tag{1.14}$$

We assume parameters $\gamma_l(T)$ and $\gamma_l(T)$ are defined by a temperature-dependence law of the form (McPherson 1986):

$$\boldsymbol{\gamma_1(T) = a_0 + a_1 \cdot k \cdot T} \tag{1.15}$$

$$\boldsymbol{\gamma_2(T) = b_0 + b_1 \cdot k \cdot T} \tag{1.16}$$

Hence, equations [1.11]–[1.16] combine to give

$$\boldsymbol{f(S_{a0}, S_{b0}, T) = (a_0 + a_1 \cdot k \cdot T) \cdot S_{a0} + (b_0 + b_1 \cdot k \cdot T) \cdot S_{b0}} \tag{1.17}$$

From equation [1.6], we get

$$\left[\frac{\partial\left(\Delta G_F^{\ddagger}(S_x,S_y,T)\right)}{\partial S_x}\right]_{S_{a0},S_{b0}} = \left[-\frac{\partial\left(f(S_x,S_y)\right)}{\partial S_x}\right]_{S_{a0},S_{b0}} = -\frac{1}{2}\cdot(a_0 + a_1 \cdot k \cdot T) \tag{1.18}$$

and $$\left[\frac{\partial\left(\Delta G_F^{\ddagger}(S_x,S_y,T)\right)}{\partial S_y}\right]_{S_{a0},S_{b0}} = \left[-\frac{\partial\left(f(S_x,S_y)\right)}{\partial S_y}\right]_{S_{a0},S_{b0}} = -\frac{1}{2}\cdot(b_0 + b_1 \cdot k \cdot T) \tag{1.19}$$

The partial derivative of equation [1.7] with respect to S_x or S_y gives

$$\left[\frac{\partial\left(\Delta H_R^{\ddagger}(S_x,S_y,T)\right)}{\partial S_x}\right]_{S_{a0},S_{b0}} = -\left[\frac{\partial\left(f(S_x,S_y)\right)}{\partial S_x}\right]_{S_{a0},S_{b0}} = -\frac{1}{2}\cdot(a_0 + a_1 \cdot k \cdot T) \tag{1.20}$$

and $$\left[\frac{\partial\left(\Delta H_R^{\ddagger}(S_x,S_y,T)\right)}{\partial S_y}\right]_{S_{a0},S_{b0}} = -\left[\frac{\partial\left(f(S_x,S_y)\right)}{\partial S_y}\right]_{S_{a0},S_{b0}} = -\frac{1}{2}\cdot(b_0 + b_1 \cdot k \cdot T) \tag{1.21}$$

The parameters a_0, a_1, b_0, b_1 and $\Delta S°$ must be determined experimentally.

Consequently, injecting equations [1.7], [1.9] and [1.10] into equations [1.6] and [1.7], we obtain:

$$\Delta G_F^{\ddagger}(S_x, S_y, T) \approx E_a - \frac{1}{2} \cdot (a_0 + a_1 \cdot k \cdot T) \cdot S_{a0} - \frac{1}{2} \cdot (b_0 + b_1 \cdot k \cdot T) \cdot S_{b0} - \frac{1}{2} \cdot (a_0 + a_1 \cdot k \cdot T) \cdot (S_x - S_{a0}) - \frac{1}{2} \cdot (b_0 + b_1 \cdot k \cdot T) \cdot (S_y - S_{b0}) \quad [1.22]$$

$$\Delta H_R^{\ddagger}(S_x, S_y, T) \approx E_a + T \cdot \Delta S° - \frac{1}{2} \cdot (a_0 + a_1 \cdot k \cdot T) \cdot S_{a0} - \frac{1}{2} \cdot (b_0 + b_1 \cdot k \cdot T) \cdot S_{b0} - \frac{1}{2} \cdot (a_0 + a_1 \cdot k \cdot T) \cdot (S_x - S_{a0}) - \frac{1}{2} \cdot (b_0 + b_1 \cdot k \cdot T) \cdot (S_y - S_{b0}) \quad [1.23]$$

Which reduce to:

$$\Delta G_F^{\ddagger}(S_x, S_y, T) \approx E_a - \frac{1}{2} \cdot (a_0 + a_1 \cdot k \cdot T) \cdot S_x - \frac{1}{2} \cdot (b_0 + b_1 \cdot k \cdot T) \cdot S_y \quad [1.24]$$

$$\Delta H_R^{\ddagger}(S_x, S_y, T) \approx E_a + T \cdot \Delta S° - \frac{1}{2} \cdot (a_0 + a_1 \cdot k \cdot T) \cdot (S_x) - \frac{1}{2} \cdot (b_0 + b_1 \cdot k \cdot T) \cdot (S_y) \quad [1.25]$$

Let us define $S_x = x \cdot S_{Br1}$ and $S_y = y \cdot S_{Br2}$ [1.26]

and $S_{a0} = x_0 \cdot S_{Br1}$ and $S_{b0} = y_0 \cdot S_{Br2}$ [1.27]

where S_{Br1} and S_{Br2} are respectively the breakdown or burnout experimental values of the two stress parameters considered for S_x and S_y. In such a case, the stresses are defined by the percentages x and y respectively for S_x and S_y with $0 < x < 100\%$ and $0 < y < 100\%$

$$\Delta G_{F_{LS}}^{\ddagger}(S_x, S_y, T) \approx E_a - \frac{1}{2} \cdot (a_0 + a_1 \cdot k \cdot T) \cdot x \cdot S_{Br1} - \frac{1}{2} \cdot (b_0 + b_1 \cdot k \cdot T) \cdot y \cdot S_{Br2} \quad [1.28]$$

$$\Delta H_{R_{LS}}^{\ddagger}(S_x, S_y, T) \approx E_a + T \cdot \Delta S° - \frac{1}{2} \cdot (a_0 + a_1 \cdot k \cdot T) \cdot x \cdot S_{Br1} - \frac{1}{2} \cdot (b_0 + b_1 \cdot k \cdot T) \cdot y \cdot S_{Br2} \quad [1.29]$$

The net reaction rate under stress conditions for the breakdown process from equation [1.1] becomes:

$$k_{net_{LS}} = exp\left(-\frac{E_a-\frac{1}{2}\cdot(a_0+a_1\cdot k\cdot T)\cdot x\cdot S_{Br1}-\frac{1}{2}\cdot(b_0+b_1\cdot k\cdot T)\cdot y\cdot S_{Br2}}{k\cdot T}\right) - exp\left(-\frac{E_a+T\cdot\Delta S^{\circ}-\frac{1}{2}\cdot(a_0+a_1\cdot k\cdot T)\cdot x\cdot S_{Br1}-\frac{1}{2}\cdot(b_0+b_1\cdot k\cdot T)\cdot y\cdot S_{Br2}}{k\cdot T}\right) \quad [1.30]$$

Hence, [1.30] expresses with a thermal term related to the activation energy as a function of stress conditions in a like-catalyst thermodynamic process and a stress term depending on the stresses applied in the form (same conclusion as in Suhir 2010):

$$k_{net} = k_1(x,y)\cdot exp\left(-\frac{E_{a_{equ}}(x,y)}{k\cdot T}\right) \quad [1.31]$$

with $\boldsymbol{E_{a_{equ}}(x,y) = E_a - \frac{1}{2}\cdot a_0\cdot x\cdot S_{Br1} - \frac{1}{2}\cdot b_0\cdot y\cdot S_{Br2}}$ [1.32]

and $\boldsymbol{k_1(x,y) = k_0\cdot exp\left(\frac{1}{2}\cdot a_1\cdot x\cdot S_{Br1} + 1/2\cdot b_1\cdot y\cdot S_{Br2}\right)}$ and

$$\boldsymbol{k_0 = \left(1 - exp\left(-\frac{\Delta S^{\circ}}{k}\right)\right)} \quad [1.33]$$

From these equations, we can observe that the **equivalent activation energy is dependent on the temperature *T_0* and is increasing with temperature**. So, under the stress conditions, **the failure mechanism model is a non-uniform acceleration mechanism**.

We can also observe that equivalent activation energies are based on the **constant Arrhenius activation energy** E_a of the mechanism under investigation but modified by a series of factors related to the stress condition applied.

The accelerating factor can be expressed considering range stress conditions. We can simply write the *AF* considering operational conditions versus reference stress conditions (sub-labeled "op", respectively "ref") as follows:

$$\boldsymbol{AF(x_{op}, y_{op}, T_{op}, x_{ref}, y_{ref}, T_{ref}) = exp\left[-\frac{E_a}{k}\left(\frac{1}{T_{op}} - \frac{1}{T_{ref}}\right)\right]}$$

$$\boldsymbol{\cdot\, exp\left[\frac{a_0\cdot S_{Br1}}{2\cdot k}\left(\frac{x_{op}}{T_{op}} - \frac{x_{ref}}{T_{ref}}\right)\right]\cdot exp\left[\frac{b_0\cdot S_{Br2}}{2\cdot k}\left(\frac{y_{op}}{T_{op}} - \frac{y_{ref}}{T_{ref}}\right)\right]}$$

$$\boldsymbol{\cdot\, exp\left[\frac{a_1}{2}\cdot S_{Br1}\cdot(x_{op} - x_{ref}) + \frac{b_1}{2}\cdot S_{Br2}\cdot(y_{op} - y_{ref})\right]} \quad [1.34]$$

1.4.1. *Precautions associated with accelerated testing*

The x and y values are assumed to range from no stress (e.g. $x = y = 0\%$) to highest stress (e.g. $x = y = 100\%$), and the AF expression is exact for any range of external stress applied. Extreme caution must be taken with these theoretical calculations. As what we use to set the acceleration is supposed to be related to failure physics without changing the physics. Indeed, let us assume a single failure mechanism of PoF – the accelerating factor calculated for various stress conditions superimposed – is non-uniform and the equivalent activation energy is stress- and temperature-dependent. This must be verified using Weibull (or log–normal) distributions when conducting proper test experiments and observing a single value of E_a activation energy parameter determined within a limited domain of range of temperature.

Based on the failure rate expression, the TTF for reference stress-accelerated conditions (REF) compared to operational conditions (labeled "op") becomes:

$$exp\left[\frac{E_a}{k}\left(\frac{1}{T_{REF}}-\frac{1}{T_{op}}\right)\right]\cdot exp\left[\frac{a_0\cdot S_{Br1}}{2\cdot k}\left(\frac{x_{op}}{T_{op}}-\frac{x_{REF}}{T_{REF}}\right)\right]\cdot exp\left[\frac{b_0\cdot S_{Br2}}{2\cdot k}\left(\frac{y_{op}}{T_{op}}-\frac{y_{REF}}{T_{REF}}\right)\right]\cdot exp\left[\frac{a_1}{2}\cdot S_{Br1}\cdot\left(x_{op}-x_{REF}\right)+\frac{b_1}{2}\cdot S_{Br2}\cdot\left(y_{op}-y_{REF}\right)\right] \quad [1.35]$$

These approaches in mathematics and physics show how the activation energy is related to the stress and temperature applied and can no longer be considered a constant to extrapolate a particular experiment under high stress to a nominal mission operation profile. This is why several end-user industries and institutions are very cautious about performing life tests so close to nominal conditions, sometimes wrongly, to new (or different) failure mechanisms while it is simply explained by the interaction of stress and temperature effects on the measured activation energy or to a best extent to Eyring law.

We may note that the burnout limits S_{Br1} and S_{Br2} are characteristic parameters that may change during aging and hence as the stress applied S_x or S_y (equation [1.27]), this means the x and y percentage stress values increase during aging from low- to high-stress levels.

The general statement that during useful life the failure rate is supposed to be constant can be wrong when approaching the end of life. When devices get closer to wearout, it is assumed that they degrade faster and their burnout limit reduces as time elapses. As such, the applied stress cannot be considered constant and reaches a high-stress level comparative to initial low stress.

This observation can provide some insight into how the failure rate can be considered non-constant and hence propose a vision on EOL of the bathtub curve.

1.5. Reliability and probability mathematics

Reliability is defined as the probability that a device will function over a certain period of time, and is usually measured in failures-in-time (FIT). FIT is a rate, defined as the number of expected device failures per 10^9 device-hours.

Failure rate (λ) is calculated by dividing the total number of failures or rejects by the cumulative time of operation. In the HTOL model, the cumulative time of operation is referred to as the number of device-hours (NDH):

$$\boldsymbol{NDH = D \cdot H \cdot A_F} \qquad [1.36]$$

where:

D = number of devices tested;

H = test hours per device;

A_F = acceleration factor derived from the Arrhenius equation and/or another stressor.

An FIT is assigned to each component multiplied by the number of devices in a system as an approximation of the expected system reliability. For the reliability model of an entire system, the FIT rates of each component in the system are summed together.

The *failure rate (λ)* in FITs (failures per billion unit-hours) is given by:

$$\boldsymbol{\lambda = \frac{\#\cdot failures}{NDH}} \qquad [1.37]$$

The conventional chi-squared expression for failure rate, λ, is:

$$\boldsymbol{\lambda = \frac{\chi^2(2n+2,1-\alpha)\cdot 10^9}{2\cdot D\cdot H\cdot A_F}} \qquad [1.38]$$

where $\frac{\chi^2(2n+2,1-\alpha)}{2}$ is the upper confidence value for "n" failures and the upper confidence limit, α (expressed in %). The value for $2n + 2$ degrees of freedom and the probability $1 - \alpha$ can be obtained from a table or calculated using Microsoft Excel functions.

According to this expression, to demonstrate zero failure rate, we must account for either a sample size up to infinity, AF up to infinity or test time up to infinity (McPherson 2019b). In such cases, we get no product to ship, the accelerating factor cannot be infinity (because this is physically impossible) or the test time will end when products are too old to ship. Achieving a zero failure rate goal using reliability testing is impossible.

A practical numerical application from standard device qualification from HTOL qualification sequence leads to the following limitation.

For defect density limits, let us assume 0 failures on a 92-device sample size (SS) under test, a 60% confidence level applied on a Poisson distribution is expressed as:

$$\boldsymbol{F} \leq \frac{\mathbf{1}}{\boldsymbol{SS}} \cdot \boldsymbol{ln}\left(\frac{\mathbf{1}}{\mathbf{1}-\boldsymbol{P}}\right) = \frac{\mathbf{1}}{\mathbf{92}} \cdot \boldsymbol{ln}\left(\frac{\mathbf{1}}{\mathbf{1}-\mathbf{0.6}}\right) = \mathbf{0.025} = \mathbf{2.5}\% \text{ or } 25{,}000 \text{ ppm.} \quad [1.39]$$

Numerical application for a study case on a FinFET technology assuming three major failure mechanisms is defined by the following Arrhenius reliability parameters (not considering other types of stressors like voltage or current for the sake of simplicity):

– EM – electromigration kinetics (Ea = 0.75eV (Al–Cu);

– NBTI kinetics (Ea = 0.6eV);

– HCI kinetics (Ea = -0.25eV long channel or +0.25 short channel).

Let us assume a qualification test experiment on 130 samples tested over a period of 2000 hrs of 150°C HTOL with 0 failures observed.

What would the verified projected operating lifetime of the device be if the operational application is at T = 45°C?

In such cases, what are the failure rates λ_i (for I – EM, NBTI or HCI) in FIT and the corresponding accelerating factors for each mechanism?

What is the TTF for each failure mechanism considering a constant failure rate based on a Poisson statistic?

What would be the equivalent total λ_{total} and corresponding TTF if all failure mechanisms are equally activated?

The hypotheses are summarized in Table 1.1a, and some answers to the questions above are given in Table 1.1b.

Qualification test conditions		
Sample Size	130	
duration	2,000	hrs
nb of hr.comp.	260,000	hr.comp.
Tstress for high temp. (EM, BTI and HCI short channel)	150	°C
Tstress for HCI long channel	0	°C
Tuse	45	°C
k (Boltzmann factor)	8.62E-05	eV/K
CL	90	%

Table 1.1a. *Example of qualification test condition for a CMOS FinFET technology with an example of associated list of failure mechanisms*

Reliability figures			**Failure mechanisms**	**AF**	**Operational time verified (in years)**	**λ Poisson (in FIT) in operation**	**MTBF (in years) 1/λ Poisson**	**CDF *Fpoisson* (@ t = 30 years)**
Ea EM	**0.75**	**eV**	Ea EM Tref = 150°C	892	204	10	11,501	0.26%
Ea NBTI	**0.6**	**eV**	Ea NBTI Tref = 150°C	229	52	39	2,955	1.01%
Ea HCI	**0.25**	**eV (short channel)**	Ea HCI (short channel) Tref = 150°C	9.6	2.2	920	124	21.48%
Ea HCI	**-0.25**	**eV (long channel)**	Ea HCI (long channel) Tref = 0°C	4.5	1.03	1969	58	40.39%
						λ total	MTBF equ.	CDF *Fpoisson* (@ t = 30 years)
						2,937	39	53.78%

Table 1.1b. *CMOS FinFET technology reliability assessment to predict operational time duration and CDF, MTTF for various failure mechanisms for CL = 90%*

From Table 1.1a, we consider many of the 130 samples used to validate a qualification test sequence, as defined by quality standards with two types of life test conditions: one at high temperature (150°C) to stress EM, BTI and HCI for short-channel transistors and one at low temperature (0°C) stress applicable to long-channel-based transistors (the HCI mechanism has a negative activation energy).

According to the life test sequence duration, the number of hours.components is 260,000.

What should the sample size and duration be to demonstrate a cumulative CDF of less than 0.1% after 30 years (as required in space applications)? Considering all failure mechanisms activated simultaneously (according to Table 1.1a), the answer is summarized in Table 1.1c and it is required to test a 200 Mhrs.comp (or 20,000 parts tested for 10,000 hrs) to guarantee the 0.1% CDF goal after 30 years at 90% CL. In such a case, the failure rate λ_{total} is 3.8 FIT, MTBF is about 29.9 Kyears and the $F_{poisson}$ (CDF) after 30 years is as low as the 0.1% goal.

To demonstrate 30 years for CDF less than 0.1%	20,000	Sample size	2E+08 hr.comp	λ_{total}	MTBF equ. (in years)	CDF *Fpoisson* (@ t = 30 years)
	10,000	hrs		3.82	29,898	0.10%

Table 1.1c. *CMOS FinFET technology. What should be the sample size and duration to demonstrate a cumulative CDF equal to less than 0.1% after 30 years with 90%CL*

The accelerating factors at 90% confidence level range broadly from 892 down to 4.5. The projected operating lifetime of the device (90% CL) verified by the qualification stress test is projected for the operational application duration at T = 45°C up to 204 years for EM mechanisms, 52 years for NBTI, 2.2 years for hot HCI and 1 year for cold HCI mechanisms. This shows how much multiple failure mechanisms can affect the demonstrated operating time projection.

The failure rate projected in operation assuming a Poisson distribution at 90% CL is assessed at 10 FIT for EM, 39 FIT for NBTI, 920 FIT for hot HCI and 1,969 FIT for cold HCI. These results are not enough to meet application reliability goals.

If supposed unconnected, each failure mechanism leads to a wide range of TTF from 1,150 years for EM, 2,955 years for NBTI, 124 years for Hot HCI and 58 years for cold HCI. Again, these data must be considered from a different point of view to be representative of the theoretical truth.

Because of this large range of values, we need to suppose that all failure mechanisms, EM, BTI and hot HCI, are equally and simultaneously activated. So, we need to cumulate their corresponding $\lambda_{Poisson}$ distribution to assess a total equivalent failure rate. This sum gives λ_{total} (90%CL) = 2.94 FIT for Poisson distribution at 90%CL. Then, the total MTTF is about 39 years for a Poisson distribution and not as supposed if we take into account the worst case of each failure mechanism.

We have to carefully consider these results because they are first extracted from a single lot testing and a reduced number of devices. The statistic helps us to predict how long a device from such a tested long sequence can be operational without any failure (still with 90% confidence level). The true λ and TTF are expected to be much greater than the values calculated. The occurrence of various failure mechanisms must be considered with some caution as their activation is strongly dependent on the conditions of test which can be far from the operational ones.

A last comment is that these failures are supposed to be constant with time (the bottom of the bathtub curve), but this is the wrong way to address the end of life reliability figure. We know that the failure mechanisms must be regarded as wearout (to the end of the bathtub curve) and we should take into account WEIBULL distribution (or LOG–NORMAL respectively) to determine the $MTTF_{wearout}$ as 63% lot failure (respectively 50% lot failure). This means that the $MTTF_{wearout}$ is not the MTTF deduced from the constant failure rate.

Therefore, HTOL is relatively poor for predicting the MTTF-associated constant failure rate and ineffective for wearout modeling as no failures are observed. Furthermore, the MTTF is not a zero-failure guarantee (it is a 50% lot failure) and we should instead define an $MTTF_{0.1\%}$ for high reliability applications.

To continue, we ask the following question: How many devices would we need to test to ensure that the fraction defective is F $\leq$35 ppm.

To answer such a question, we need to make use of the "stress resistance" concept. Implementation of this method usually refers to information referred to as "lump sum". The most basic version of the stress resistance method calculates the probability of failure which results from the interaction between two probability laws. In general, if any two laws are explicitly known, the probability of failure can be calculated by the numerical resolution of a convolution integral. Only a few particular cases lead to purely analytical solutions, when the two laws belong to the same statistical family. Among the most classic examples, we can cite the interaction between pairs of homologous laws, normal or log–normal, which allow us to obtain exact analytical solutions, as presented in Pierrat (2017) and Delaux

(2019). Nevertheless, assuming the correct hypotheses, approximation can be set (Pierrat and Delaux 2013). At the 60% confidence level (P = 0.6) and with accepting on finding zero defects (x = 0), the sample size (SS) is deduced from various statistical distributions:

$$\text{Poisson: } SS = \frac{1}{F} \cdot ln\left(\frac{1}{1-P}\right) = \frac{1}{35ppm} \cdot ln\left(\frac{1}{1-0.6}\right) = 26{,}180\ devices$$

$$\text{Chi square: } SS = \frac{[\chi^2(P,2)]/2}{F} = \frac{[\chi^2(0.6,2)]/2}{35\ ppm} = \frac{1.8326/2}{35\ ppm} = 26{,}180\ devices$$

So, for the above sample size, SS > 50, little or no differences are observed in statistical distributions. In fact, such situations are not entirely realistic because, in general, the laws resulting from statistical estimates deviate from the theoretical laws to which they are assimilated.

The acceleration factor, AF, is the most problematic consideration. The industry recognizes that an AF is given only for each failure mechanism individually by specific testing from the manufacturer, and is defined as the ratio between the time it takes a certain fraction of the devices to fail, under a certain amount of stress or use conditions, and the corresponding time under more severe stress or use conditions. In the case of multiple mechanisms, the TTF for each mechanism must be considered separately for each associated AF per mechanism.

The TTF of each mechanism can also be defined from the physical models of the mechanisms, described in the JEDEC publication, JED-122G (JEDEC 2011) (see also McPherson 2019a):

TDDB anode hole injection exponential 1/E model (Chen et al. 1985) from the Fowler–Nordheim tunneling current:

$$\boldsymbol{TF_{TDDB_{\frac{1}{E}model}}} = e^{G(T)\cdot\frac{1}{E_{0x}}} \cdot e^{\frac{Ea_{TDDB}}{k\cdot T}} \qquad [1.40]$$

TDDB thermoelectrical exponential E model (McPherson and Baglee 1985):

$$\boldsymbol{TF_{TDDB_{Emodel}}} = e^{\gamma(T)\cdot E_{ox}} \cdot e^{\frac{Ea_{TDDB}}{k\cdot T}} \qquad [1.41]$$

TDDB power law voltage V-model for hyper-thin Si02 dielectrics (<40Å):

$$\boldsymbol{TF_{TDDB_{Power\ lawmodel}}} = \boldsymbol{B_0(T)\cdot V_{ox}}^{-n} \cdot e^{\frac{Ea_{TDDB}}{k\cdot T}} \qquad [1.42]$$

TDDB exponential $E^{1/2}$ model:

$$TF_{TDDB_{E1/2model}} = C_0(T) \cdot e^{-\alpha \cdot \sqrt{E}} \cdot e^{\frac{Ea_{TDDB}}{k \cdot T}} \quad [1.43]$$

Hot carrier injection (Takeda and Suzuki 1983):

$$AT = f \cdot \left(\frac{I_{sub}}{W}\right)^{-N} \cdot e^{\frac{Ea_{HCI}}{k \cdot T}} \quad [1.44]$$

Negative-bias temperature instability:

$$TF_{NBTI} = V^{\frac{\alpha}{m}} \cdot e^{\frac{Ea_{NBTI}}{k \cdot T}} \quad [1.45]$$

Electromigration (Black 1969):

$$TF_{EM} = A_0 \cdot (J - J_{crit})^{-p} \cdot e^{\frac{Ea_{EM}}{k \cdot T}} \quad [1.46]$$

Stress migration (McPherson and Dunn 1987) due to mechanical stress inducing the plastic deformation of metal with time

SM in aluminum or copper interconnects:

$$Creep(voiding)rate = B_0 \cdot (T_O - T)^n \cdot e^{\frac{Ea_{SM_{Al}}}{k \cdot T}} \quad [1.47]$$

Humidity model (HM) exponential reciprocal humidity:

$$TF_{HM_{exp-reciprocal}} = A_0 \cdot e^{\frac{b}{RH}} \cdot e^{\frac{Ea_{HM}}{k \cdot T}} \quad [1.48]$$

HM power law HM:

$$TF_{HM_{power-law}} = A_0 \cdot RH^{-n} \cdot e^{\frac{Ea_{HM}}{k \cdot T}} \quad [1.49]$$

HM exponential HM:

$$TF_{HM_{power-law}} = A_0 \cdot e^{-a \cdot RH} \cdot e^{\frac{Ea_{HM}}{k \cdot T}} \quad [1.50]$$

Thermal-cycling/fatigue-induced mechanisms:

– The Coffin–Manson model for low-cycle fatigue

$$\#Cycle - to - Failure = A_0 \cdot \left(\Delta\varepsilon_p\right)^{-s} \quad [1.51]$$

where $\Delta\varepsilon_p$ is the plastic strain range.

– The Coffin–Manson model for temperature cycling

$$\Delta\varepsilon_p \propto (\Delta T - \Delta T_0)^{\beta} \quad [1.52]$$

Failure rates are often expressed in terms of failure units (FITs): $1FIT = 1$ failure in 10^9 device-hours.

TTF is statistical, and a distribution of TTF is observed. Life distributions are defined based on three mathematical functions:

a) the probability density function for device or system failures *f(t)* which relates the relative frequency of failure to time, *t*;

b) the cumulative distribution function, *F(t)*, which gives the summation of failure proportion to that time;

c) the curve of failure rate $\lambda(t)$.

The instantaneous failure rate, the hazard rate function $\lambda(t)$, is the ratio of the number of failures during the time period Δt for the devices that were healthy at the beginning of testing (operation) to the time period Δt

$$\lambda(t) = \frac{f(t)}{1-F(t)} \quad [1.53]$$

The cumulative probability distribution function (CDF) *F(t)* for the probability of failure is related to the probability density distribution function *f(t)* as

$$F(t) = \int_0^t f(x) \cdot dx \quad [1.54]$$

and the reliability function *R(t)*, the probability of non-failure is defined as

$$R(t) = 1 - F(t) \quad [1.55]$$

Probability data obtained when performing accelerated tests can be modeled by various distribution models such as the exponential law, Weibull law, normal or log–normal distributions, beta distributions, etc. The main life distribution functions are:

– exponential;

– normal;

– lognormal;

– Weibull distribution.

1.5.1. *Exponential distribution summary*

Probability density function: $f(t) = \lambda \cdot e^{-\lambda \cdot t}$ [1.56]

Cumulative density function: $F(t) = 1 - e^{-\lambda \cdot t}$ [1.57]

Instantaneous failure rate, hazard rate: $\lambda(t) = \lambda$ [1.58]

Significant properties

Mean (average) of MTBF: $MTBF = 1/\lambda$ [1.59]

1.5.2. *Normal distribution summary*

Probability density function: $f(t) = \frac{1}{\sigma \cdot \sqrt{2\pi}} \cdot e^{-\frac{1}{2}\left(\frac{t-\mu}{\sigma}\right)^2}$ [1.60]

Cumulative density function: $F(t) = \frac{1}{\sigma \cdot \sqrt{2\pi}} \cdot \int_0^t e^{-\frac{1}{2}\left(\frac{x-\mu}{\sigma}\right)^2} \cdot dx$ [1.61]

Instantaneous failure rate, hazard rate: $\lambda(t) = \frac{\frac{1}{\sigma \cdot \sqrt{2\pi}} \cdot e^{-\frac{1}{2}\left(\frac{t-\mu}{\sigma}\right)^2}}{1 - \frac{1}{\sigma \cdot \sqrt{2\pi}} \cdot \int_0^t e^{-\frac{1}{2}\left(\frac{x-\mu}{\sigma}\right)^2} \cdot dx}$ [1.62]

Significant distribution properties

Median (50% failure): $t = t_{50\%} = \mu$

Mean (average): $t = \mu$

Location parameter: μ

Shape parameter: σ

s, estimate of σ can be calculated as $t_{50\%} - t_{16\%}$

1.5.3. *Weibull distribution summary*

Probability density function: $f(t) = \frac{\beta \cdot (t-\gamma)^{\beta-1}}{\alpha^\beta} \cdot e^{-\left(\frac{t-\gamma}{\sigma}\right)^\beta}$ [1.63]

Cumulative density function: $F(t) = 1 - e^{-\left(\frac{t-\gamma}{\sigma}\right)^\beta}$ [1.64]

Instantaneous failure rate, hazard rate: $\lambda(t) = \frac{\beta \cdot (t-\gamma)^{\beta-1}}{\alpha^{\beta}}$ [1.65]

Significant properties:

Location parameter (63% failure): $t_{63\%} = \alpha$

Shape parameter: β

Time-delay parameter = γ, a factor used only when the data do not fit the distribution, except with the use of a time delay.

1.5.4. *Lognormal distribution summary*

Probability density function: $f(t) = \frac{1}{\sigma \cdot t \cdot \sqrt{2\pi}} \cdot e^{-\frac{1}{2} \cdot \left(\frac{\ln t - \mu}{\sigma}\right)^2}$ [1.66]

Cumulative density function: $F(t) = \frac{1}{\sigma \cdot \sqrt{2\pi}} \cdot \int_0^t \frac{1}{x} \cdot e^{-\frac{1}{2} \cdot \left(\frac{\ln x - \mu}{\sigma}\right)^2} \cdot dx$ [1.67]

Instantaneous failure rate, hazard rate: $\lambda(t) = \frac{\frac{1}{\sigma \cdot t \cdot \sqrt{2\pi}} \cdot e^{-\frac{1}{2} \cdot \left(\frac{\ln t - \mu}{\sigma}\right)^2}}{1 - \frac{1}{\sigma \cdot \sqrt{2\pi}} \cdot \int_0^t \frac{1}{x} \cdot e^{-\frac{1}{2} \cdot \left(\frac{\ln x - \mu}{\sigma}\right)^2} \cdot dx}$ [1.68]

Significant properties:

Median (50% failure): $t = t_{50\%} = e^{\mu}$

Mean (average): $t - e^{\mu + \frac{\sigma^2}{2}}$

Location parameter: e^{μ}

Shape parameter: σ

s, estimate of σ can be calculated as $\ln \frac{t_{50\%}}{t_{16\%}}$

Their schematic representation behaviors are given in Figure 1.12. When considering devices or system life, three failure types can be distinguished and are commonly represented by the bathtub curve.

Failures that appear in the early stages of component life are called early (infantile) failures. They can be explained by faulty manufacturing and insufficient quality control during production. They can be eliminated by a systematic screening test. However, some component weaknesses can pass through this filter, either because the test conditions are not demanding enough to effectively screen for latent defects or because the failure criterion set is defined to eliminate catastrophic failures and a certain level of drift failures.

Random failures, the second category, cannot be eliminated by a screening test or by an optimal use policy (maintenance). They can be provoked by sudden voltage increases that can strongly influence the component quality and reliability. These failures appear erratically, accidentally, unforeseeably. In a paper by Ed Sperling, editor in chief of Semiconductor Engineering (Sperling 2019), which included interviews with car manufacturers, the following was noted by Gert Jørgensen, vice president of marketing at Delta Microelectronics: "You can't test for the random failures." Sperling continues:

> [...] the problem is worse in automotive due to both the expected lifespan of systems and the harsh environmental conditions [...] True random failures are rare. A stray alpha particle hitting a circuit and causing damage is known to happen, and the chances of that occurring increase with denser circuits and thinner insulation. So a single-event upset affecting a 7 nm device with FinFETs packed tightly together is more likely than at 28 nm. The same is true for random contaminants, which may affect one part differently than another. But delineating which failures are truly random from those that are not is time-consuming, and that adds to the cost and slows down time to market.

Wearout failures, the third category, are an indicator of component aging.

Two methods are generally used to make reliability estimates: (i) the parts count method and (ii) the parts stress analysis method. The parts count method requires less information, generally that dealing with the quantity of different part types, the quality level of the parts and the operational environment.

Parts stress analysis requires the greatest amount of detail and is applicable in the later design phases where actual hardware and circuits are being designed.

Whichever method is used, the objective is to obtain a reliability estimate that is expressed as a failure rate. Calculation of failure rate for an electronic assembly, unit or system requires knowledge on the failure rate of each part contained in the item of interest. If we assume that the item will fail when any (all in series) of its parts fail, the failure rate of the item will equal the sum of the failure rate of its parts.

Parts count reliability prediction requires the following information (DoD 1991; FIDES 2004): (i) generic part types (including complexity of microelectronics) and quantities; (ii) part quality levels; and (iii) equipment environment. The general expression for equipment failure rate with this method is for the given environmental conditions:

$$\lambda = \sum_{i=1}^{n} N_i \cdot (\lambda_G \cdot \pi_Q)_i \qquad [1.69]$$

λ = total equipment failure rate (in FIT = 1 failure in 10^9 hrs);

λ_G = generic failure rate for the ith generic part;

π_Q = quality factor for the ith generic part;

N_i = quantity of the ith generic part;

n = number of different generic parts.

In the parts stress analysis method, part failure models vary with different part types, but their general form is:

$$\lambda_i - \lambda_B \cdot \pi_E \cdot \pi_A \cdot \pi_Q \cdots \pi_N \qquad [1.70]$$

λ_B = reference failure rate without environmental stress;

π_E = environmental factors other than temperature stressors (e.g. external stressors such as vibration and humidity);

π_A = adjustment factors, for example, biasing conditions (either static or dynamic or signal stressors, etc.);

π_Q = quality factors referring to the quality control applied during manufacturing, screening and lot assurance testing before shipment to the user.

It should be noted that a series of multiplicative factors assumes the effect of each stressor can be counted independently. This appears to be somewhat of a contradiction because the effects are not really independent, so we express the failure rate $\lambda(x,y)$ from equation [1.35] as:

$$\lambda = \lambda_0 \cdot exp\left(\frac{E_a}{k \cdot T}\right) \cdot exp\left[-\frac{a_0 \cdot S_{Br1} \cdot x}{2 \cdot k \cdot T}\right] \cdot exp\left[-\frac{b_0 \cdot S_{Br2} \cdot y}{2 \cdot k \cdot T}\right] \cdot exp\left(\frac{-a_1}{2} \cdot S_{Br1} \cdot x\right) \cdot exp\left(\frac{-b_1}{2} \cdot S_{Br2} \cdot y\right) \qquad [1.71]$$

with

$$\lambda_0 = \frac{1}{k_0} = \left(1 - exp\left(-\frac{\Delta S^\circ}{k}\right)\right)^{-1} \quad [1.72]$$

Life distributions	Probability density function $f(t)$	Reliability function $R(t)$	Hazard Failure Rate $\lambda(t)$
Normal	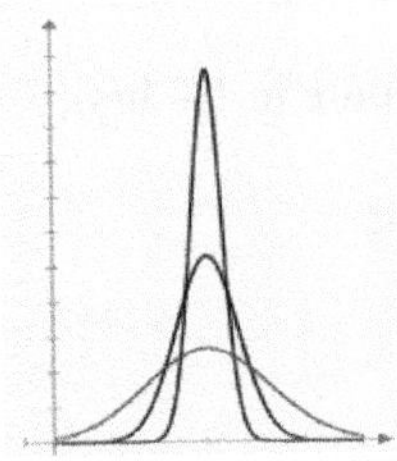	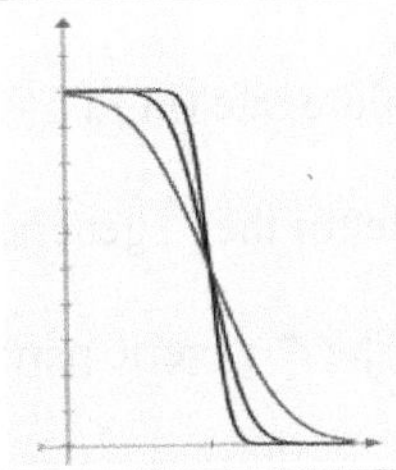	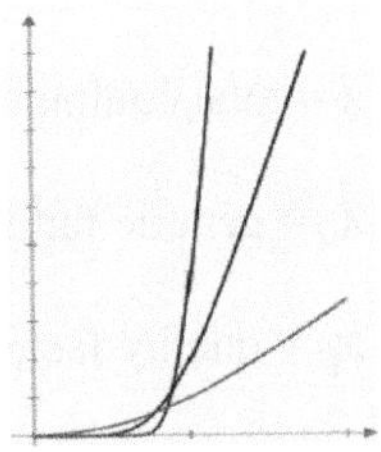
Exponential	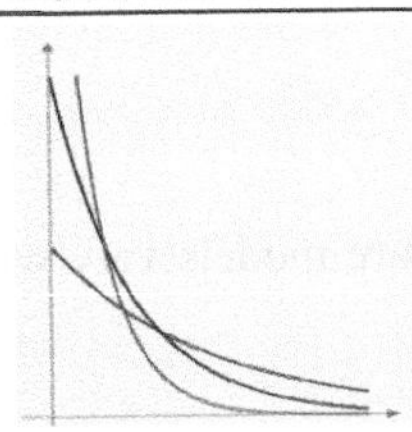	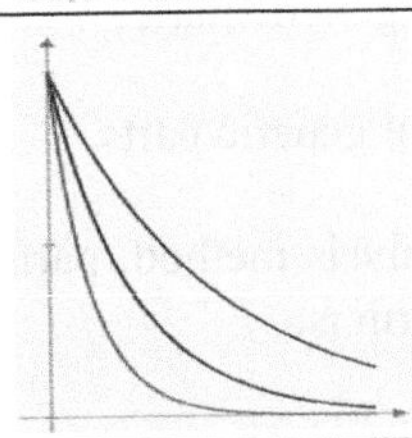	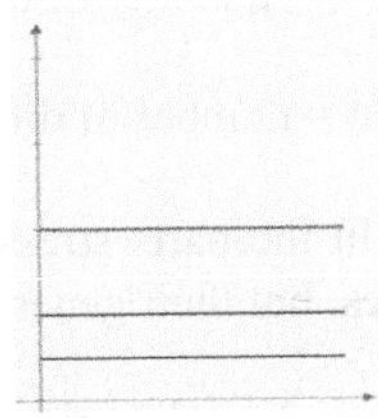
Log–normal	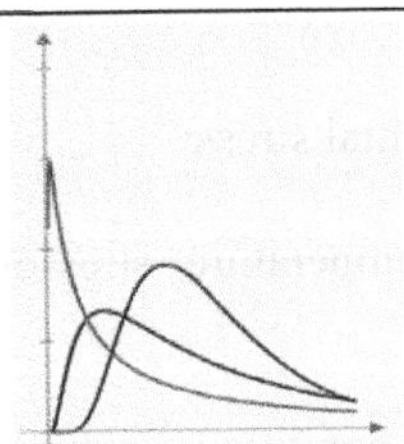	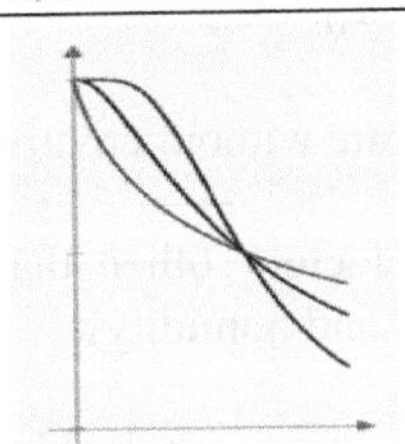	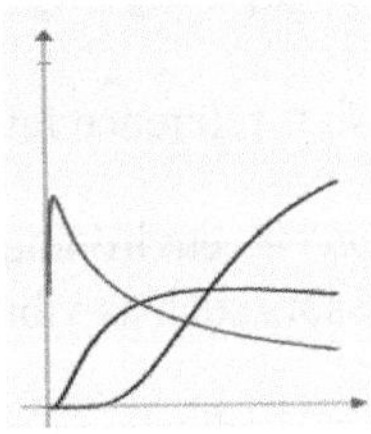
Weibull	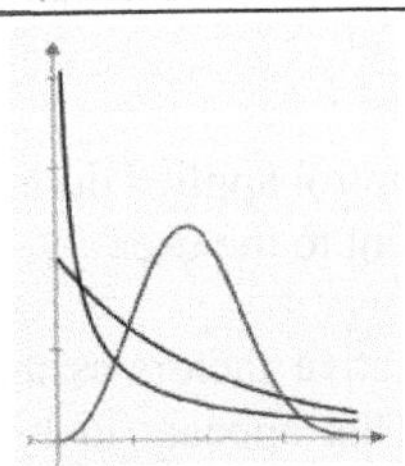	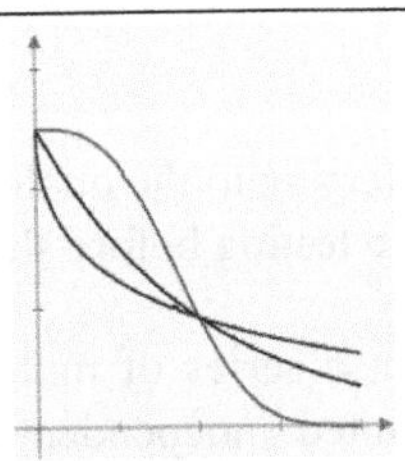	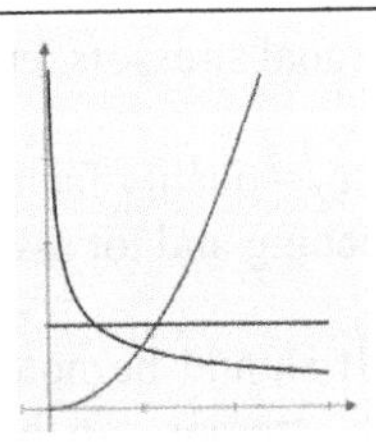

Figure 1.12. *Life distribution functions represented schematically. For a color version of this figure, see www.iste.co.uk/elhami/mechatronics.zip*

Comparing equations [1.35] and [1.70], we observe that the π factors can be defined as:

a) those listed in (DoD 1991) or (FIDES 2004)

$$\pi_{T,T_{REF}} = exp\left[\frac{E_a}{k}\left(\frac{1}{T_{REF}} - \frac{1}{T_{op}}\right)\right] \quad [1.73]$$

$$\pi_{S_{1op},S_{1REF}} = exp\left[\frac{a_1}{2} \cdot S_{Br1} \cdot \left(x_{op} - x_{REF}\right)\right] \quad [1.74]$$

$$\pi_{S_{2op},S_{1REF}} = exp\left[\frac{b_1}{2} \cdot S_{Br2} \cdot \left(y_{op} - y_{REF}\right)\right] \quad [1.75]$$

those not considered in (DoD 1991) or (FIDES 2004)

$$\pi_{S_{1op},S_{REF},T_{op},T_{REF}} = exp\left[\frac{a_0 \cdot S_{Br1}}{2 \cdot k}\left(\frac{x_{op}}{T_{op}} - \frac{x_{REF}}{T_{REF}}\right)\right] \quad [1.76]$$

$$\pi_{S_{2op},S_{REF},T_{op},T_{REF}} = exp\left[\frac{b_0 \cdot S_{Br2}}{2 \cdot k}\left(\frac{y_{op}}{T_{op}} - \frac{y_{REF}}{T_{REF}}\right)\right] \quad [1.77]$$

From this, we observe that the equivalence with actual standards is incomplete because of missing terms, those related to external environmental stressors such as vibration, mechanical, radiation, humidity, etc.

1.6. The Sedyakin principle

It would be very useful if we could somehow convert a rather complicated dynamical stress pulse, over some time interval t_1–t_2, into a rectangular pulse effective stress which would produce an equivalent amount of material/device degradation over this same time interval t_1–t_2.

Assuming an accelerated failure time (AFT) model (e.g. only the time scale is affected under stress), in both cases, accelerating factors (AF) can apply based on [1.34]. Quality standards also define the similar rate function models and related AF. In 1966, Sedyakin (1966) formulated his famous physical principle in reliability which states that for two identical populations of units functioning under different stresses S_1 and S_2, two moments t_1 and t_2 are equivalent if the probabilities of survival until these moments are equal. Figure 1.13 shows how this principle also relates to the resulting reliability function $R(t)$ for a population submitted to a step stress condition (these can be generalized to a superposition of various stressors like temperature, voltage, etc.).

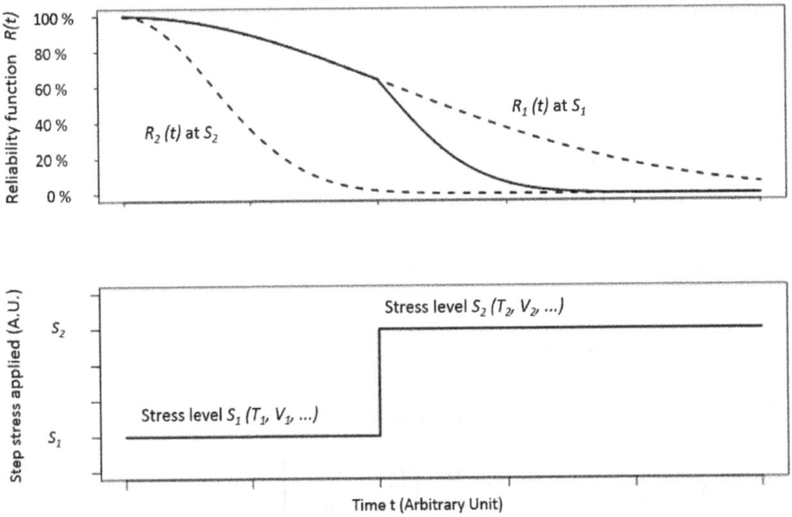

Figure 1.13. *Reliability function R(t) as a function of step stress applied and the Sedyakin principle*

As presented by McPherson (2019a) (Chapter 14), and potentially calling on the Sedyakin principle, the TTF models developed assume that the stress remains constant with time until the product fails. In semiconductor products (transistors, integrated circuits or any active electronic device), the applied signals induce voltage, current and power stressors that change during operation and are generally frequency-dependent. So, it is convenient to convert dynamical (time-dependent) stress to an equivalent effective static stress. Basic mathematics to determine the effective stress $\xi_{effective}$ such that it produces an equivalent amount of degradation and thus the same TTF as the dynamical stress $\xi(t)$ is developed in McPherson (2019a) using the following concept of the **compliance equation** for periodic (period P) dynamical stress and equivalent to power-law TF models:

$$\frac{1}{P} \cdot \int_0^P AF_{\xi(t),\xi_{effective}} \cdot dt = 1 \quad [1.78]$$

where the accelerating factor for the power-law TF model is for a yield stress $\xi_{yield} < \xi(t)$:

$$AF_{\xi(t),\xi_{effective}} = \left(\frac{\xi(t) - \xi_{yield}}{\xi_{effective} - \xi_{yield}}\right)^n \quad [1.79]$$

Gives:

$$\xi_{effective} - \xi_{yield} = \left[\frac{1}{P} \cdot \int_0^P \left(\xi(t) - \xi_{yield}\right)^n \cdot dt\right]^{\frac{1}{n}} \qquad [1.80]$$

It is also interesting to consider how to calculate the effective static temperature equivalent proposed by McPherson. He mentioned that similar to stress, the temperature T of a device is not constant but depends on the biasing on–off operation, pulse operation, thermal cycling, as well as on the dynamic signal applied.

To assess the effective static temperature $T_{effective}$ which produces an equivalent amount of device degradation versus the temperature variation $T(t)$, the compliance equation to determine $T_{effective}$ over a time interval of $(t_a - t_b)$:

$$\frac{1}{(t_a - t_b)} \cdot \int_0^P AF_{T(t),T_{effective}} \cdot dt = 1 \qquad [1.81]$$

1.7. System reliability

1.7.1. *Series systems*

How do we estimate the reliability distribution parameters and failure probabilities for complex DSM components operating under harsh conditions, as applied during their mission profiles?

The newest component technologies deploy transistors with very small gate lengths (less than 10 nm) with new materials, reduced dimensions and new technologies. Once the emerging technologies for numerical applications under high environmental constraints for true and real-time mission profiles are validated, the issue is not yet resolved.

In Chapter II, we have identified CMOS bulk reliability distribution parameters accounting for electrical aging mechanisms, such as high-temperature degradation, so-called negative or positive thermal instability (NBTI, PBTI) or hot carrier injection (HCI) and the hard or soft breakdown of dielectric gates. These mechanisms can cause significant reductions in performance and software errors that are very important to quantify under nominal operating conditions. We need to determine their impact on the lifetime of the single sensitive active area of a device and then look at how it can degrade the performance of the design. It is important to note that most of these mechanisms do not lead directly to a straightforward failure but only to a gradual parametric drift of the elementary functions which will only induce a failure when the parameters exceed a critical threshold strongly dependent on the architecture. For the dominant mechanisms in recent technologies, this failure rate increases over time, a characteristic property of wearout mechanisms. Moreover, the reliability of the current circuits often leads to no failures being observed in these mechanisms for the duration of the tests (a few months). This does not make it possible to evaluate the probability of failure beyond the equivalent duration of the test; this probability is then likely to increase significantly because of the increasing nature of the failure rate.

How can these probabilities be used to predict the total component DSM system performance? How does the DSM design affect reliability? How does redundancy design in a DSM affect total reliability?

What are the hypotheses to consider a system reliability model like a DSM?

How component failure rate characteristics are described: multiple failure mechanisms superimposed, dependent or independent, constant (cataleptic failures) vs non-constant. What if a failure results in an open path or in a short circuit path and what could the impact be on series or parallel system modeling?

This chapter derives the formula for the reliability of a series system, parallel system or a mixed series/parallel system, as fully detailed in Tobias and Trindade (2012).

The most commonly used model for system reliability assumes that the system is made up of n independent components which must all operate in order for the system to function properly. But this series model is applied specifically when a single integrated circuit with several independent failure modes is analogous to a system with several independent elementary constituents. The failure mechanisms are competing with each other in the sense that the first to reach a failure state causes the component to fail: the open question is still to consider what a failure state is. Is it for catastrophic failures or related to a failure criteria or performance? As argued by Tobias and Trindade (2012): "the more general competing risk model where the failure processes for different mechanisms are not independent can be very complicated since one must know how the random times of failure for different mechanisms are correlated".

Assuming the first hypothesis, and the ith element have a cumulative distribution function (CDF) $F_i(t)$, the probability that the series system fails at time t is the probability that one or more of the n independent components have failed at time t. In terms of total probability, we consider the product of the n CDFs:

$$R_s\,(t) = Pr(E) = \Pr(E_1, E_2, \dots, E_n) = \Pr(E_1) \cdot \Pr(E_2) \cdot \dots \cdot \Pr(E_n) = \prod_{i=1}^{n} R_i(t) \qquad [1.82]$$

with

$$R_i(t) = e^{-\lambda_i \cdot t} \quad [1.83]$$

where λ_i is the failure rate of the *ith* component. Hence,

$$\lambda_s(t) = \sum_{i=1}^{n} \lambda_i \quad [1.84]$$

If the components have the same reliability R_{s0}, meaning $\lambda_i = \lambda_0$ for $i = 1$ to n, this becomes:

$$R_s(t) = R_{s0}^n = \left(e^{-\lambda_0 \cdot t}\right)^n = e^{-n \cdot \lambda_0 \cdot t} \quad [1.85]$$

LEMMA 1.1.– A series system constituted of n identical and independent elements, each described by a Poisson distribution (λ_0) reliability model can be approximated by a general equivalent Poisson distribution with parameter $n \cdot \lambda_0$.

1.7.2. *Parallel systems*

A system that operates with n elements in parallel survives until the last of its element fails, considering in this case a failure mode definition to be an open circuit. We will not consider the second hypothesis case if the failure mode of each element is a short circuit. In such a case, the system with n elements in parallel survives if only one element fails.

Assuming the first hypothesis, and the ith element have a cumulative distribution function (CDF) $F_i(t)$, the probability that the parallel system fails at time t is the probability that all the components have failed at time t. In terms of total probability, we consider the product of n CDFs:

$$F_p(t) = \prod_{i=1}^{n} F_i(t) \quad [1.86]$$

$$R_p(t) = 1 - \prod_{i=1}^{n} [1 - R_i(t)] \quad [1.87]$$

1.7.2.1. *Poisson distribution function*

If the CDF is defined by Poisson or the exponential distribution function, the probability the system fails, for example, at least one of the n elements fails, is the reliability or probability of success at time t expressed as:

$$R_p(t) = 1 - \prod_{i=1}^{n} \left(1 - e^{-\lambda_i \cdot t}\right) \quad [1.88]$$

If $\lambda_i = \lambda$ for any $i = 1$ to n

$$R_p(t) = 1 - \left(1 - e^{-\lambda \cdot t}\right)^n \qquad [1.89]$$

Considering the life mission t is defined by the plateau of the bathtub, for $\lambda \cdot t << 1$, we can approximate the reliability at time t by:

$$e^{-\lambda \cdot t} \approx 1 - \lambda \cdot t \qquad \text{or} \qquad \lambda \cdot t \approx 1 - e^{-\lambda \cdot t} \qquad [1.90]$$

and merging [1.89] with [1.90] gives:

$$R_p(t) \approx 1 - (\lambda \cdot t)^N \approx e^{-(\lambda \cdot t)^n} = e^{-\left(t/\theta\right)^\beta} \qquad [1.91]$$

The error generated by the approximation above is assessed and plotted in Figure 1.14, showing [1.89] and [1.91] versus parameter $\lambda{\cdot}t$ in units of MTTF. The error generated by the approximation is less than 1% for $\lambda \cdot t \leq 0.3$ or, in other words, at time t lower than or equal to 30% of MTTF (for $\lambda = 100$ FIT and n = 3, MTTF = $1/\lambda = 10^7$ hrs). This error decreases for higher values of n.

In fact, such a parallel system on n independent and identical elements having the same constant Poisson failure rate λ (e.g. same equivalent $MTTF_i = MTTF = 1/\lambda$ for $i = 1$ to n) shows an equivalent Weibull distribution for short times $t \leq 30\% \cdot$MTTF range for an error lower than 1%.

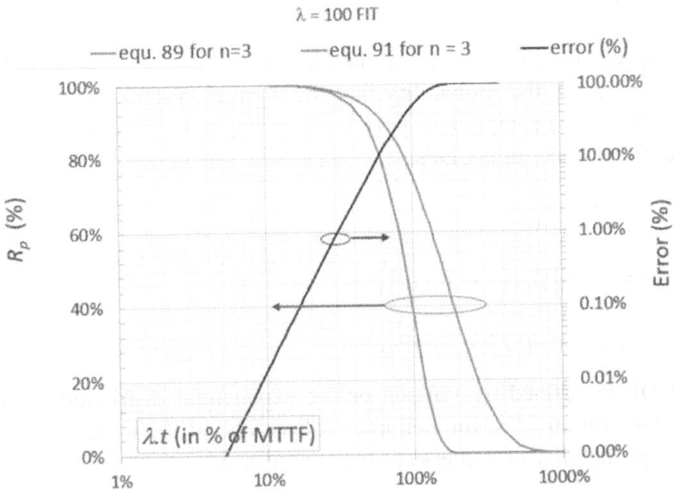

Figure 1.14. *Plot error (in red) as the difference between the plot of equations [1.89] and [1.91] for λ = 100 FIT and n = 3. For a color version of this figure, see www.iste.co.uk/elhami/mechatronics.zip*

LEMMA 1.2.– A parallel system constituted of n identical and independent elements, each described by a Poisson distribution (λ) reliability model can be approximated by a general equivalent Weibull distribution with parameters λ and $\beta = n$ with an error lower than 1% for time operation lower than 30%·MTTF for the example shown.

1.7.2.2. *Weibull distribution function*

The random variable T has a Weibull distribution with parameters α_i and β_i if the reliability or probability of success at time t is given by:

$$R_{iW}\,(t, \alpha_i, \beta_i) = e^{-\left(\frac{t}{\alpha_i}\right)^{\beta_i}}, \qquad t \geq 0 \qquad [1.92]$$

$$R_{pW}\,(t, \alpha_i, \beta_i) = 1 - \prod_{i=1}^{n}\left(1 - e^{-(\beta_i \cdot t^{\mu_i})}\right) \qquad [1.93]$$

For identical independent elements, we set $\alpha_i = \alpha$ and $\beta_i = \beta$ for any $i = 1$ to n, so:

$$R_{pW}\,(t, \alpha, \beta) = 1 - \left[1 - e^{-\left(\frac{t}{\alpha_i}\right)^{\beta_i}}\right]^n \qquad [1.94]$$

Considering the life mission t is defined by the plateau of the bathtub, for $\lambda \cdot t << 1$, and we can get for a parallel system and short time t:

$$R_{pW}\,(t, \alpha, \beta) \approx e^{\;({}^{t}/_{\alpha})^{\beta \cdot n}} \qquad [1.95]$$

The error generated by the approximation of equation [1.95] is assessed and plotted in Figure 1.15, showing equations [1.94] and [1.95] versus parameter $\lambda{\cdot}t$. The error generated by the approximation is less than 1% for $\lambda \cdot t \leq\ 0.3$ or, in other words, at time t lower than or equal to 80% of MTTF (e.g. λ = 100 FIT and n = 5). This error decreases for higher values of n and β.

LEMMA 1.3.– A parallel system constituted of n identical and independent elements, each described by a reliability Weibull distribution (α, β) model can be approximated by a general equivalent Weibull distribution with parameter (α, β·n) with an error lower than 1% for time operation lower than 75%·MTTF.

As a major fact and a generalization point of view, such a parallel system based on n independent and identical elements having the same constant Weibull failure rate parameters (α, β) is shown to be modeled by an equivalent Weibull distribution with reliability parameter *(α, β·n)*. The error induced by such an approximation is estimated for short times t ≤ 80%·MTTF and range for an error lower than 1% for the example shown for $\lambda = 100$ FIT, $\beta = 5$ and $n = 3$.

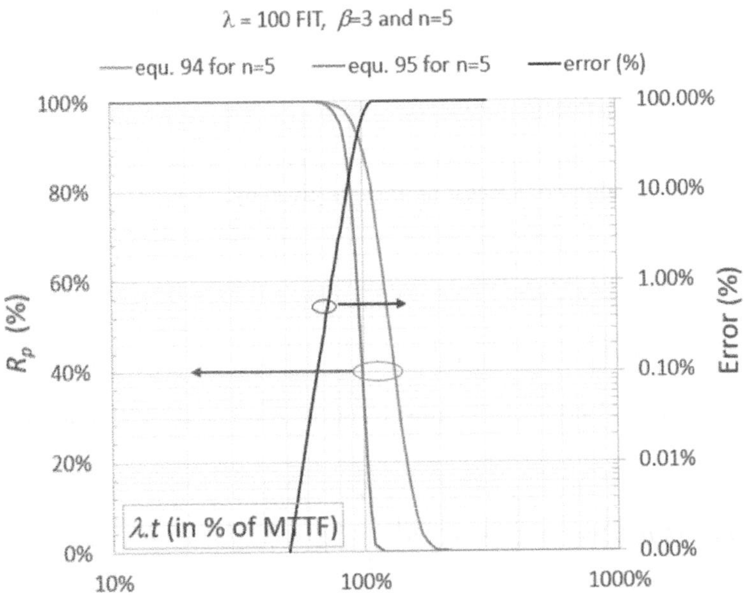

Figure 1.15. *Plot error (in red) as the difference between equations [1.94] and [1.95] for $\lambda = 100$ FIT, $\beta = 3$ and n = 5. For a color version of this figure, see www.iste.co.uk/elhami/mechatronics.zip*

1.7.3. *Complex systems*

Physical configurations in series or parallel do not necessarily indicate the same logic relations in terms of reliability. An integrated circuit (DSM) composed of LUTs connected in logical paths contains billions of elementary interconnected structures. From a reliability perspective, the LUT includes multiple transistors in series-parallel configurations but an LUT is said to have failed if one or more transistors have failed. So, the transistors in a LUT are considered in series from a reliability perspective.

There are systems that require more than one component to succeed in order for the entire system to operate. In addition, the performance of a product is usually measured by multiple characteristics. In many applications, there is one critical characteristic, which describes the dominant degradation process. This can be used to characterize product reliability. The failure of a product can be defined in terms of performance characteristics or indicators that cross a specific threshold. Figure 1.16

depicts the relationship between the degradation path, pseudo-life and life distribution. The maximum likelihood method should be used for the estimation of distribution parameters. Once the reliability parameters have been estimated, we can use a Monte Carlo simulation to generate a large number of degradation paths. The probability of failure *F(t)* is approximated by the percentage of simulated degradation paths crossing a specified threshold after a stress time applied.

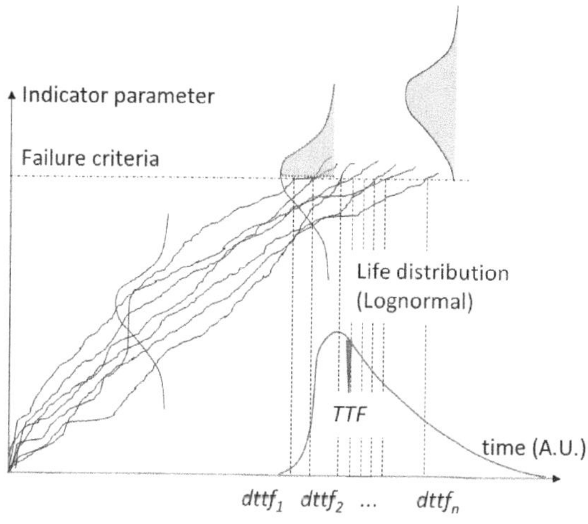

Figure 1.16. *Relation of degradation path, pseudo-life and life distributions. For a color version of this figure, see www.iste.co.uk/elhami/mechatronics.zip*

When considering the catastrophic failure paradigm, a general series system can be observed in a configuration for a DSM FPGA test structure based on the ring oscillator (RINGO) test structure represented in Figure 1.17.

Nevertheless, if we consider the parametric failure mechanism hypothesis, another representation should be proposed, as shown in Figure 1.18. In such a case, a single component with four independent parametric failure mechanisms (HCI, BIT, EM or TDDB) occurring simultaneously is analogous to a series system. Each failure mechanism "competes" with the others to cause a failure, and failure rates are additive, mechanism by mechanism, and so drifts are cumulated. In a RINGO structure, a failure is observed when all NAND fail in a cumulated drift failure reaching the failure criteria. In such a case, the reliability model is similar to a parallel system because by definition only the failure of all components within the system results in the failure of the entire system. In other words, we know a parallel system succeeds if one or more components are operational. Considering a parametric drift failure, the NANDs work together pending to reach the cumulated parameter drift limit because the contribution of each failure mechanism is accumulated.

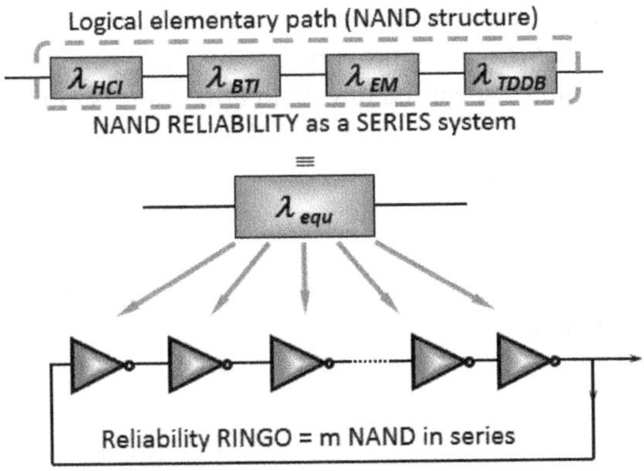

Figure 1.17. *Series reliability configuration for DSM RINGO catastrophic failure mode modeling (four failure mechanisms and m = RINGO size defined as the number of NAND interconnected). For a color version of this figure, see www.iste.co.uk/elhami/mechatronics.zip*

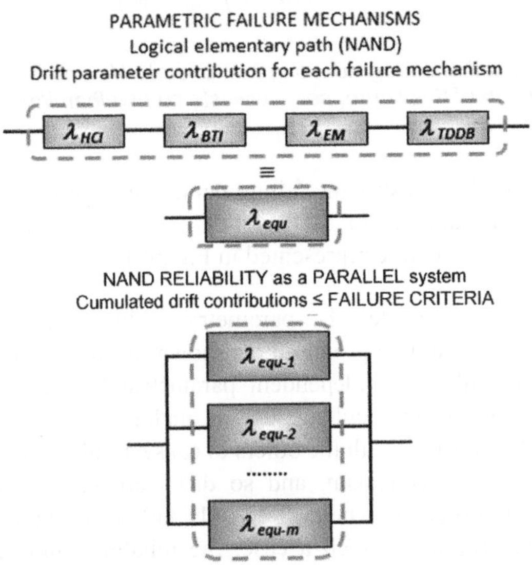

Figure 1.18. *Series-parallel reliability configuration for DSM RINGO parametric failure mode modeling (m = RINGO size defined as the number of NAND interconnected). For a color version of this figure, see www.iste.co.uk/elhami/mechatronics.zip*

The reliability of a series-parallel system defined in Figure 1.18 and for a ***Poisson distribution function*** is written as:

$$R_{p-sp}\left(t,\alpha_{i,j},\beta_{i,j}\right) = 1 - \prod_{j=1}^{m}\left(1 - \prod_{i=1}^{4} e^{-\lambda_i \cdot t}\right) \quad [1.96]$$

Considering m NAND elements in parallel, each with four types of failure mechanisms in series, we get

$$R_{p-sp}(t) = 1 - \prod_{j=1}^{m}\left(1 - e^{-t\cdot\sum_{i=1}^{4}\lambda_i}\right) \quad [1.97]$$

Because all NAND elements are similar, it can be written as:

$$R_{p-sp}(t) = 1 - \left[1 - e^{-t\cdot\sum_{i=1}^{4}\lambda_i}\right]^m \quad [1.98]$$

If $\sum_{i=1}^{4}\lambda_i \cdot t << 1$, for example, assuming true up to a few percent of MTTF, we can approximate the reliability at time t by:

$$\left(\mathbf{1} - \boldsymbol{e}^{-\sum_{i=1}^{4}\lambda_i \cdot t}\right)^{\boldsymbol{m}} \approx \left(\boldsymbol{t}\cdot\sum_{i=1}^{4}\boldsymbol{\lambda}_i\right)^{\boldsymbol{m}} \quad [1.99]$$

And equation [1.98] reduces to:

$$R_{p-sp}(t) \approx 1 - \left(t\cdot\sum_{i=1}^{4}\lambda_i\right)^m \quad [1.100]$$

Allowing us to express:

$$R_{p-sp}(t) \approx 1 - \left(t\cdot\sum_{i=1}^{4}\lambda_i\right)^m \approx e^{-\left(t\cdot\sum_{i=1}^{4}\lambda_i\right)^m} = e^{-\left(t/\theta\right)^\beta} \quad [1.101]$$

as a Weibull general equivalent representation, where

$$\frac{1}{\theta} = \sum_{i=1}^{4}\lambda_i \quad \text{and} \quad \beta = \text{m} \quad [1.102]$$

which is indeed the number of NANDs incorporated in a RINGO test structure.

LEMMA 1.4.– In a series-parallel system constituted of n identical and independent elements, each described by a reliability Poisson distribution (λ), its reliability model can be approximated by a general equivalent Weibull distribution with parameter (θ, β= m) with an error lower than 3% for time operation lower than 50%·MTTF (see Figure 1.19).

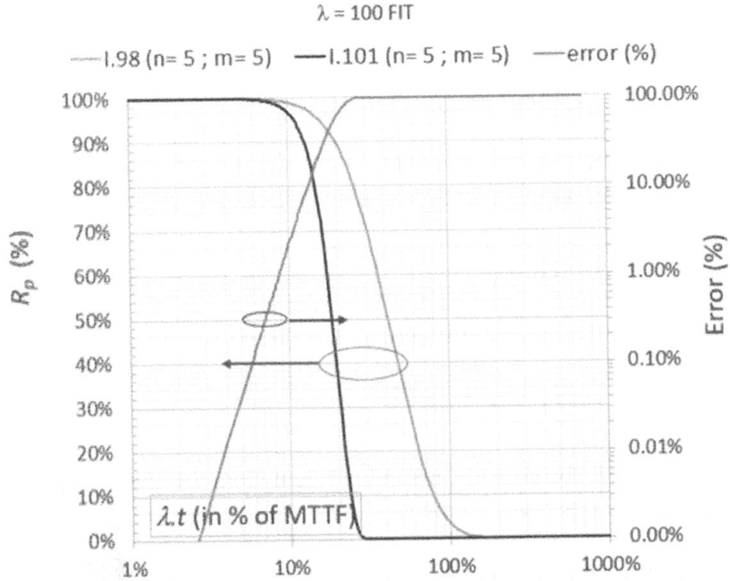

Figure 1.19. *Plot error (in red) as the difference between the plot of equations [1.98] and [1.101] for a Poisson distribution with λ = 100 FIT, β = m = 5. For a color version of this figure, see www.iste.co.uk/elhami/mechatronics.zip*

1.8. Conclusion and future prospects

Using emerging microelectronic COTS appears to be necessary to achieve today's product and system performances, with the drawback of inadequate quality standards that do not properly address how to qualify them in an appropriately short period of time. Indeed, at the end of the 20th century, the introduction of a new component in a space system required several years of evaluation for complete qualification. Such timing was compatible with technology evolution and the satellite development and manufacturing phases. Today, with the explosion of satellite constellations and their short development and production phase lasting only two to three years, such standards are not adapted in some cases.

To counter this drawback, we need to better understand the physics of semiconductors and failure mechanisms in order to quickly identify and select "reliable COTS" technology and lot screening before authorizing their implementation in complex systems. By "reliable COTS", we consider we have enough information consolidated through short-term adaptive experiments to identify how we can guarantee the mission profile as existing for low Earth orbit (LEO) satellite constellations or long-term geostationary Earth orbit telecommunication satellites. The same type of considerations must be fixed based on skills and knowledge of physics of the devices to draw optimized experiments and deep analyses related to the automotive or aeronautic industries.

A reliability engineering team needs to bring together multi-skilled people with valuable knowledge on (this list is not exhaustive):

– electronic design, to be able to understand how the devices are biased (DC and digital or analog signals) and stressed from normal to extreme mission profile conditions;

– semiconductor and assembly process manufacturing, characterization and metrology techniques, to understand how devices are packaged and tested in order to determine and validate their safe operating area limits and to define a sequence of screenings for efficient freak distribution purging, and their ability to survive environmental and application conditions during operating life;

– physics of semiconductors, physics of new microelectronic devices, induced failure modes and mechanisms to understand how devices are designed and manufactured to determine the key electrical stressors and indicators to monitor degradation patterns influenced by internal and external stress conditions (respectively biasing and environmental);

– reliability concepts and modeling, to identify and consolidate hypotheses of degradation based on a series of experiments (as short as possible) representative of true mission conditions.

In the paradigm of the physics of failure, people focus on how a product is failing either catastrophically or gradually.

In the PoH, we propose to focus on the characterization and physical reliability knowledge of lot devices able to assure the system mission in operation before failure. The goal is to provide advice to the equipment engineer on associated quality and reliability risks.

The very short development of new technologies into the custom market has seized upon microelectronic COTS products, and consequently led to a change in the way high reliability applications are designed.

Chapter 1 has been dedicated to understanding how the PoH concept can be defined. To do so, we first described two types of emerging DSM technologies (very narrow node size, for example, < 10 nm and GaN power DC switch transistor) in a short state-of-the-art review. Setting aside the random failure of devices, system or equipment breakdowns generally due to handling, packaging or external overstress, we have focused on the description of wearout mechanisms. Taking advantage of

McPherson's *Reliability Physics and Engineering* (McPherson 2019a), we presented and detailed the mathematics based on the Gibbs free energy diagram, considering reactions to several external stresses applied and temperature. The multi-failure mechanisms are also considered to develop a reliability model and to establish an accelerating factor multi-stress and multi-mechanism expression based on non-constant activation energy.

Reliability models were also recalled with respect to random failure rate modeling, and system reliability was presented from the formula of series systems, parallel systems or a mixed series/parallel systems, as detailed in Tobias and Trindade (2012).

Establishing and recalling this preliminary set of mathematics, we propose to develop Chapter 2 to show how these tools can be applied and used experimentally to predict pre-failure occurrence and elaborate predictive reliability SHM protocols. A set of sequences are described to help diagnose PoH capability on new COTS devices through a series of control steps also named the Prognostics and Health Management Protocol. Such a protocol can be easily applied both on new, emerging and existing technologies.

1.9. References

Agarwala, B., Attardo, M.J., Ingraham, A.P. (1970). Dependence of electromigration-induced failure time on length and width of aluminium thin-film conductors. *Journal of Applied Physics*, 41, 3954.

Arrhenius, S. (1896). Über den Einfluss des Atmosphärischen Kohlensäurengehalts auf die Temperatur der Erdoberfläche. *Proceedings of the Royal Swedish Academy of Science*, 26(1), 6–11.

BELLCORE (2011). SR-332, Reliability Prediction Procedure for Electronic Equipment. BELLCORE Telcordia.

Bensoussan, A. and Suhir, E. (2013). Design-for-reliability (DfR) of aerospace electronics: Attributes and challenges. *IEEE Aerospace Conference*, 2–9 March.

Bensoussan, A., Coval, P., Roesch, W., Rubalcava, T. (1994). Reliability of a GaAs MMIC process based on 0.5µm Au/Pd/Ti gate MESFETs. *32nd Annual Proceeding Reliability Physics, IRPS*, San Jose, CA, April.

Bensoussan, A., Durier, A., Bernstein, J.B. (2017). M-Storm: A new approach to model electronics components reliability. *Automotive Power Electronics (APE 2017)*, Paris.

Bernstein, J.B. (2015). Reliability prediction for aerospace electronics. *IEEE Aerospace Conference*, Big Sky, MN.

Bernstein, J.B. and Bensoussan, A. (2016). Reliability prediction with MTOL, proposed to IEEE system qualification for reliability. *Transaction on Devices and Material Reliability (TDMR)*.

Black, J. (1969). Black, electromigration – A brief survey and some recent results. *IEEE Transactions on Electron Devices*, ED-16, 388.

Boltzmann, L. (1886). Boltzmann, the second law of thermodynamics. Populare Schriften. Essay 3, Address to a formal meeting, Imperial Academy of Science.

Bravaix, A. (2019). Part 2 – DDR3 FF20 Reliability Analysis – ISEN. Internal report, IRT-ISEN, Toulouse, September.

Chen, I.C., Holland, S., Hut, C. (1985). A quantitative physical model for time-dependent breakdown in SiO2. *23rd International Reliability Physics Symposium Conference*, 25–29 March.

Cox, D. (1972). Regression models and life-tables. *Journal of the Royal Statistical Society. Series B (Methodological)*, 34(2), 187–220.

Cox, D. and Oakes, D. (1984). *Analysis of Survival Data*, Volume 21. CRC Press, Boca Raton, FL.

Creveling, C. (1956). Increasing the reliability of electronic equipment by the use of redundant circuits. *Proc. IRE*, 44, 509–515.

Decker, K. (1994). GaAs MMIC hydrogen degradation study. *GaAs Reliability Workshop*, Philadelphia, PA.

Delaney, M., Wiltsey, T., Chiang, M., Yu, K. (1994). Reliability of 0.25µm GaAs MESFET MMIC process: Results of accelerated lifetests and hydrogen exposure. *GaAs Reliability Workshop*, Philadelphia, PA.

Delaux, D. (2019). Fiablité appliquée pour l'industrie. Internal report, Valeo PTS, Paris.

Dhillon, B. and Rayapati, S. (1986). A method to evaluate reliability of three-state device network. *Microelectronics Reliability Journal*, 26, 525–554.

Drenick, R. (1960). Mathematical aspects of the reliability problem. *Journal of the Society for Industrial and Applied Mathematics*, 8(11), 125–149.

Evans, M. and Polanyi, M. (1938). Inertia and driving force of chemical reaction. *Faraday Society (London) Trans.*, 34, 11–29.

Eyring, H., Lin, S., Lin, S. (1980). *Basic Chemical Kinetics*. John Wiley & Sons, New York.

FIDES (2004). A reliability methodology for electronic systems. Guide, DGA.

Glasstone, S., Laider, K.J., Eyring, H. (1941). *The Theory of Rate Processs: The Kinetics of Chemical Reaction, Viscosity, Diffusion and Electrochemical Phenomena*. McGraw-Hill, New York.

Guo, A. and Del Alamo, J. (2015). Positive-bias temperature instability (PBTI) of GaN MOSFETs. *IEEE International Reliability Physics Symposium*, Monterey, CA.

Hammond, G. (1955). A correlation of reaction rates. *Journal of the American Chemical Society*, 77, 334–338.

Huard, V., Cacho, F., Federspiel, X. (2012). Technology scaling and reliability challenges in the multicore era. *IRPS*, 16–18 April.

INTEL (2002). *VDSM Issues and Design Methodology*. Intel Microelectronics Services, January.

JEDEC (2017). JESD22-A108 HTOL temperature, bias, and operating life. July [Online]. Available at: www.jedec.org.

JESD85 (2021). Methods for Calculating Failure Rates in Units of FITs. JEDEC Solid State Technology, Arlington, VA.

Jin, D. and Del Alamo, J. (2012). Impact of high-power stress on dynamic ON-resistance of high-volatge GaN HEMTs. *Microelectronics Reliability Journal*, 52, 2875–2879.

Jones, S. (2019). LithoVision 2019 – Semiconductor technology trends and their impact on lithography [Online]. Available at: https://semiwiki.com/semiconductor-services/ic-knowledge/8022-lithovision-2019-semiconductor-technology-trends-and-their-impact-on-lithography/ [Accessed 28 February 2019].

Kayali, S., Ponchak, G., Shaw, R. (1996). *GaAs MMIC Reliability Assurance Guidelines for Space Applications*, Volume 96-25. JPL Publication, Pasadena, CA.

Lapedus, M. (2018). Big Trouble At 3nm [Online]. Available at: https://semiengineering.com/big-trouble-at-3nm/ [Accessed 28 February 2019].

Lee, C., Welch, B., Fleming, W. (1981). Reliability of AuGe/Pt and AuGe/Ni ohmic contacts on GaAs. *Electronics Letters*, 17, 407–408.

Liu, C., Sagong, H.-S., Kim, H., Choo, S., Lee, H., Kim, Y. (2015). Systematical study of 14 nm FinFET reliability: From device level stress to product HTOL. *IRPS*, Monterey, CA.

Loubet, N., Hook, T., Montanini, P., Yeung, C.-W., Kanakasabapathy, S., Guillorn, M., Yamashita, T., Zhang, J., Miao, X., Wang, J. et al. (2017). Stacked nanosheet gate-all-around transistor to enable scaling beyond FinFET – T17-5. *Symposium on VLSI Technology Digest of Technical Papers*, Kyoto.

McPherson, J.W. (1986). Stress dependent activation energy. *Proc. IEEE 24th IEEE IRPS*, 1–3 April.

McPherson, J.W. (2019a). *Reliability Physics and Engineering – Time-to-Failure Modeling*, 3rd edition. Springer Nature Switzerland AG, Plano, TX.

McPherson, J.W. (2019b). Future of reliability testing – Can zero failures become reality? *ESREF 2019*, Toulouse, September.

McPherson, J.W. and Baglee, D. (1985). Acceleration factors for thin gate oxide stressing. *Proc. 23rd Annual IEEE IRPS*, 25–29 March.

McPherson, J.W. and Dunn, C.F. (1987). A model for stress-induced metal notching and voiding in very large-scale-integrated Al–Si (1%) metallization. *Journal of Vacuum Science & Technology B: Microelectronics Processing and Phenomena*, 5, 1321 [Online]. Available at: https://doi.org/10.1116/1.583609.

McPherson, J.W., Khamankar, R., Shanware, A. (2000). Complementary model for intrinsic TDDB in SiO2 dielectrics. *Journal of Semiconductor Science and Technology*, 15, 462.

Meneghesso, G., Verzellesi, G., Danesin, F., Rampazzo, F., Zanon, F., Meneghini, M., Zanoni, E. (2008). Reliability of GaN high-electron-mobility transistors: State of the art and perspectives. *IEEE Transactions on Device and Materials Reliability*, 8(2), 332–343.

Meneghesso, G., Meneghini, M., Tazzoli, A., Ronchi, N., Stocco, A., Chini, A., Zanoni, E. (2010). Reliability issues of gallium nitride HEMTs. *International Journal of Microwave and Wireless Technologies*, 2(1), 39–50.

Meneghini, M., Hilt, O., Fleury, C., Silvestri, R., Capriotti, M., Strasser, G., Pogany, D., Bahat-Treidel, E., Brunner, F., Knauer, A. et al. (2016). Normally-off GaN-HEMTs with p-type gate: Off-state degradation, forward gate stress and ESD failure. *Microelectronics Reliability Journal*, 58, 177–184.

Meneghini, M., Meneghesso, G., Zanoni, E. (2017). Power GaN devices – Materials, applications and reliability. In *Power Electronics and Power Systems*, Chow, J.H., Stankovic, A.M., Hill, D. (eds). Springer, Berlin.

Nikulin, M., Couallier, V., Gerville-Reache, L. (2009). Accelerated life and degradation models in reliability and safety: An engineering perspective. *EA 2961, Statistique Mathématique et ses Applications*, Univ. Bordeaux 2, Bordeaux.

Pierrat, L. (2017). Estimation and validation of predictive reliability. Technical report, LJC-VAL/PTS, LJ-CONSULTING.

Pierrat, L. and Delaux, D. (2013). Analytical improvement of the stress-strength method by considering a realistic strength distribution. *Applied Reliability Symposium*, Berlin.

QUANTERION (n.d.). 217PlusTM:2015. Handbook of reliability prediction models [Online]. Available at: http://www.theriac.org.

Quay, R. (2008). *Gallium Nitride Electronics*. Springer, Berlin/Heidelberg.

RDF (2000). (UTE C 80-810, IEC-62380-TR Ed.1) [Online]. Available at: http://www.ute-fr.com/La-normalisation/UTE-and-standardisation.

Sedyakin, N. (1966). On one physical principle in reliability theory. *Techn. Cybernetics*, 80–87.

Serma Technologies (1996). Alcatel Espace Internal failure analysis report EL 4103. Report, Serma Technologies, Toulouse, March.

SIEMENS AG (1986). SN 29500. Reliability and quality specifications failure rates of components. Siemens Technical Liaison and Standardisation.

Snow, E., Grove, A., Deal, B., Sah, C. (1965). Ion transport phenomena in insulating films. *Journal of Applied Physics*, 36, 1664.

Sperling, E. (2019). Gaps emerge in automotive test [Online]. Available at: https://semiengineering.com/gaps-emerge-in-automotive-test/ [Accessed 08 May 2019].

Srinivasan, J. and Adce, S.V. (2003). The case for microarchitectural awareness of lfetime reliability. Technical report, UIUC CS n° UIUCDCS-R-2003-2391.

Stathis, J., Mahapatra, S., Grasser, T. (2018). Controversial issues in negative bias temperature instability. *Microelectronics Reliability Journal*, 81, 244–251.

Suhir, E. (2010). Probabilistic design for reliability. *ChipScale Reviews*, 14(6).

Suhir, E. (2013). Predicted reliability of aerospace electronics: Application of two advanced concepts. *IEEE Aerospace Conference*, 2–9 March.

Takeda, E. and Suzuki, N. (1983). An empirical model for device degradation due to hot-carrier injection. *IEEE Electron Device Letters*, EDL-4, 111–113.

Tobias, P.A. and Trindade, D.C. (2012). *Applied Reliability*, 3rd edition, CRC Press, Boca Raton, FL.

Ueda, T. (2014). Reliability issues in GaN and SiC power devices. *IEEE International Reliability Physics Symposium (IRPS)*, Waikoloa, Hawaii.

Verkest, D. (2018). Imec roadmap [Online]. Available at: https://community.cadence.com/cadence_blogs_8/b/breakfast-bytes/posts/imec2 [Accessed 28 February 2019].

Wamock, S. and Del Alamo, J. (2017). OFF-state TDDB in high-voltage GaN MIS-HEMTs. *IEEE International Reliability Physics Symposium (IRPS)*, Monterey, CA.

Wigner, E. (1938). The transition state method. *Faraday Society (London) Trans.*, 34, 29–41.

Wu, Y., Chen, C.-Y., Del Alamo, J. (2014). Temperature-accelerated degradation of GaN HEMTs under high-power stress: Activation energy of drain-current degradation. *ROCS*, 19 May.

Yunm, H., Kahirdeh, A., Christine, M., Modarres, M. (2018). Entropic approach to measure damage with applications to fatigue. *Annual Reliability and Maintainability Symposium (RAMS)*, Reno, NV, 1–6.

Zhurkov, S. (1965). Kinetic concept of the strength of solids. *International Journal of Fracture Mechanics*, 1(4), 311–323.

2

Applied Engineering on Physics-of-Healthy and SHM of Microelectronic Equipment for Aeronautic, Space, Automotive and Transport Operations

Experience without theory is blind, but theory without experience is mere intellectual play.

Immanuel KANT (1724–1804)

2.1. Introduction

In reliability standards, the MTTF is related to random mechanisms which are supposed to reach failure sooner than the wearout MTTF. This is mainly because the considered technologies are rather mature and the applied mission stress conditions are low enough to push the limits of the wearout events.

Several end-user industries and institutions are very cautious about performing life tests so close to nominal conditions. Last year, automotive OEMs began demanding that electronic components last 18 years with zero failures.

Nevertheless, new technologies and innovative components have shown possible shorter lifetime occurrence (Condra et al. 2005) because of their immaturity. When procured as COTS but used in long-term mission applications or high-rel equipment, it is especially necessary to perform screening to eliminate freak distribution and run

---

Chapter written by Alain BENSOUSSAN, Joseph B. BERNSTEIN and Alain BRAVAIX.

plenty of qualification customized programs adapted to the foreseen application. It is not easy and the simplest way is to ask the component manufacturer to help to quantify these parameters. Sometimes, manufacturers may provide information about the lifetime, which is generally based on cumulative failures observed during burn-in and qualification test sequences (considered rather as random defects).

The concept of stressor parameters is introduced to quantify how the stress will impact the equivalent activation energy described in Part I for the principle of the BAZ model (Suhir 2013) and the energy band diagram (McPherson 1986). Indeed, the external stress (any kind: electrical, signal, or external as thermal cycling or radiation, for example) generally (but not always) reduces the height of the thermal activation energy measured from pure thermal stress conditions.

The Physics-of-Healthy (PoH) concept: we define stressor and predictor/indicator electrical parameters:

– A stressor parameter is a setting condition applied to the component of interest which stresses and ages the device. Such a parameter can either be an electrical life test at a nominal setting (outlined in the percentage of the limit of the capability of the component defined as in the maximum rating conditions: thermal, DC bias, AC or RF signal) or an external environment stress condition (thermal shock, thermal cycling, radiation, ionic contamination or gas poisoning).

– A predictor (or indicator) is an in situ measurable and quantifiable electrical parameter used to assess the healthiness of a component introduced in equipment. Such a component is supposed to drive the reliability of the final card installed in the equipment. To do so, the card should be outfitted with a dedicated electrical point of access or accessible through the bias access in order to be monitored during operation (see, for example, DSM ASIC based on 10 nm node described in the chapter). The aim is to quantify how the internal characteristics and performances are affected during the time that the stress is applied.

Chapter 2 presents context and particular issues for automotive application using emerging technologies. We describe the methodology implemented by the GIFAS consortium (2015) (*Groupement des Industries Françaises Aéronautiques et Spatiales*) and the FIDES group (FIDES 2004). The predictive reliability and health monitoring methodology concept for immature technologies are deeply considered by the GIFAS consortium. We show the three-loop methodology and the schematics of the main steps involved in the reliability determination of DDR3 (FF20) and FPGA (FF16) CMOS technology nodes. We take advantage of the prognostic failure model (PFM) level 3 recommendations and determine how to apply reliability prediction on DSM technologies in harsh environments.

Failure mode and effect analysis (FMEA) is a method for reliability analysis that tries to assess and identify existing failure modes of a system. When applying the standards method at the component level, the failure mode rate is considered as a constant. Furthermore, its value is an estimation depending on the database used or extracted from few experiments. The more realistic approach proposed by the FIDES group is based on reliability prediction models, but considers one mechanism at a time. They have shown that the failure mode rate is not a constant, but depends on mission profile conditions. The FIDES tool developed considers the failure rate as stress-dependent and can be expressed by COX models.

When multiple failure mechanisms co-exist, the standard and FIDES methodology must be revisited.

We present a case study on application for automotive and aerospace. It affords a series of numerical applications supported by experiments for study cases on a FinFET technology, assuming HCI (Hot Carrier Injection), BTI (Bias Temperature Instability) and EM (Electromigration) as major failure mechanisms. It shows how the activation energy is related to the stress and temperature applied and proof that the activation energy can no longer be considered as constant to extrapolate some experiment under high stress to nominal mission operation.

In section 2.2, we give highlights (advantages and drawbacks when compared to existing standards (MTOL)) and recommendations when implementing a true case study on DRAM IC DDR3. In particular, how to consider the occurrence of multiple failure mechanisms on a component used under operational real harsh environment profiles. For the aerospace electronics reliability, we relate a pragmatic application of MTOL based on Bernstein's approach. The prediction of system reliability can be described using a linear matrix solution.

In section 2.2.2.1, we develop the PFM background.

How do we build a methodology to forecast the reliability of a device, board or system in harsh operating conditions?

What are the clues and what evidence is there to correctly and objectively answer if there is a risk or not when operating new technology, in a repeatable, safe and precise analysis process?

How do we automatize such a method in order to reduce uncertainty and avoid the influence of many expert positions, or even give a single expert the right to return to their analysis years later to improve it using new evidence and data?

How do we discriminate the concepts of constant vs non-constant rates when people are locking a pragmatic solution to assess the risk of failure during the entire mission life of a piece of equipment, product or component?

A reliability risk analysis of the lifetime of an electronic device versus the mission profile of the final product is defined as a dedicated PFM guideline. The method proposes four levels, depending on the data and service available from the manufacturer. Level 1 is the simplest to implement, and level 4 is the most complex.

2.2. Component health monitoring: a case study for automotive and aerospace applications

2.2.1. *Context and particular issues for automotive applications using emerging technologies*

Several end-user industries and institutions are very cautious about performing life tests so close to nominal conditions. Last year, automotive OEMs began demanding that electronic components last 18 years with zero failures. If we consider a car manufacturer reporting about 7,000 semiconductor devices in its high-end models, and it turns out about 4,000 cars every day, then the failure in the range of one part per million equals 24 defective cars per day. For another manufacturer that uses about half as many electronic components but manufactures 10,000 cars every day, that equates to 54 defective cars.

"Normally it takes five to six years before you know if there is a problem," according to Jim McLeish, senior quality/reliability consultant manager at DFR Solutions. "With 5 nm to 7 nm technology, there is no experience. We don't know what the variation will be." The use of 10/7 nm chips in automotive applications is making testing more challenging. There are many more transistors in each area. This requires higher test coverage, and a lot of testers are in the beginning phase of bringing technology up to an acceptable level.

It is not always clear what is a killer defect and what is not. Some latent defects may never cause failures, while others that are less obvious can develop into more serious issues under various environmental conditions or excessive electrical surge current. Figure 2.1.a shows an example of MMIC MESFET technology in the mid-1990s (Roesch 2000), where micro-particles were generated during the integration of MIM capacitor manufacturing. Based upon inputs from customers, the failure rate was estimated to be about 200 defectives per million devices. At that rate, we would have expected a 50% chance of seeing three capacitor failures in the 15,000 devices that were life-tested between 1985 and 1995 – but we did not experience any (probability was not helping us). Over the course of two years, the particle density had been reduced by the MMIC manufacturer by almost two orders of magnitude (see Figure 2.1.b) – still without generating a single accelerated reliability test failure! The size of these micro-defects ranges below 1 μm diameter, but when considering the dielectric thickness of 0.2 μm as a conformal deposition process, some of them generated early failure short circuits, while a small portion remained latent defects. Safe operating conditions were demonstrated and validated because we conducted 100% screening applying maximum voltage (destroying killer defects) prior to releasing devices for high-reliability applications.

a) Capacitor with a microparticle (1 μm diameter) on the bottom electrode (courtesy Alcatel Espace: Serma Technologies Constructional Analysis report (Serma Technologies 1996))

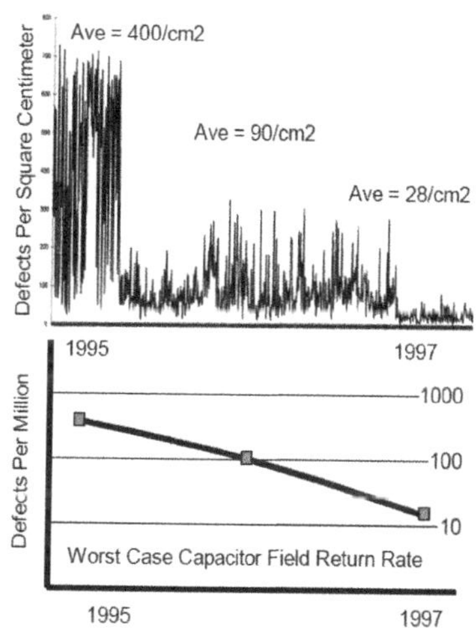

b) Defects and failure rates of capacitors (Roesch 2000)

Figure 2.1. *Example of an MMIC MESFET technology in the mid-1990 (Roesch 2000)*

Current predictive reliability modeling tools such as MIL, 217[+] or FIDES are based on Arrhenius' law, which expresses the lifetime to failure (time-to-failure or TTF) as a function of an invariable activation energy and temperature. It considers only one failure mechanism and temperature as a major stress. These tools are limited to predicting the constant failure rate.

The TTF assessment depends on whether we consider failure mechanisms to be related either to random failure or to wearout failure mechanisms. For the emerging DSM technology based on nanometric scale, the TTF calculation strongly depends on such overlapping properties between random and wearout mechanisms, as we will see in section 2.2.3. In addition to this confusion, the activation energy is not a true constant in the temperature range and mission biasing conditions applied, as we will illustrate in section 2.2.4.

A predictive reliability paradigm should consider these limitations and propose how to:

– handle the multi-failure mechanisms that can occur simultaneously either at random or at end of life;

– assign correct activation energy values to random and wearout failure mechanisms;

– express activation energy in terms of temperature and operating biasing condition ranges;

– deduce extrapolation of TTF under validated hypotheses (be careful to design accelerating stress conditions that are adequate for the true operating mission profile (e.g. mitigate and validate the representativeness of failure mechanisms activated under high stress)).

During environmental tests, we need to demonstrate that the probability of failure of a part is less than a given value, which leads to showing, with n tests, that the mean of the resistance (particular quantile) has (at least) a given value. With n tests, the average should be demonstrated to be within a certain confidence interval with a given level of confidence.

2.2.2. *Predictive reliability and health monitoring methodology for new technologies*

The lifetime of a component and thus of an electronic card is quantified using a constant random failure rate (λ). This law of probability applies to rare events or situations where the manifestation of an event occurs randomly over time with a low probability. The reliability estimation problem turns out to be different if we consider a product subjected to a degradation process that can be quantified over time and on which a "criterion of acceptability" can be set. It is then possible to track the evolution of this drift during stress tests in order to evaluate the forecast reliability of the product, or during its operational life, to know its state of health from day to day, and estimate its remaining potential.

In reliability standards, the MTTF is related to random mechanisms that are supposed to be reached sooner than the wearout MTTF. This is mainly because the considered technologies are rather mature and the applied mission stress conditions are low enough to push the limits of the wearout events.

Nevertheless, new technologies and innovative components (by definition less mature), have shown possible shorter lifetime occurrence (Condra et al. 2005).

Consequently, in parallel to the constant failure rate evaluation, it must be verified that the wearout of a component will occur far later than the lifetime determined by the random failure law. So, the concern when predicting reliability and implementing health monitoring on emerging technologies is distinguishing between random and wearout failure mechanisms. It is not easy, and the simplest way is to ask the component manufacturer to help to quantify these parameters. Sometimes, manufacturers may provide information about lifetime, generally based on cumulative failures observed during burn-in and qualification test sequences (considered instead as random defects).

The definition of MTTF (either random or wearout) is becoming a major issue and must be well distinguished as each of them assess its own value. Several major hypotheses must be described and considered:

– Failure mechanisms are based on physics of failure (PoF) models in terms of failure rate (FR) and accelerating factors (AF) compared to a reference set of stress conditions.

– The standard model of accelerated life applies, which assumes that only the scale parameter of the reliability distribution is modified by the level of stress and that the shape of the failure distribution remains unchanged according to the following relationship, and as shown in Figure 2.2:

$$R_{stress}(t) = R_{mission}(AF \cdot t) \quad [2.1]$$

– The modeling approach assumes that the same physical mechanisms can cause early "random" failures and intrinsic wearout, but act on different parts of a strength distribution (e.g. abnormal/defective sites (Bechtold et al. 2014)).

– Wearout assessments are expressed in terms of "time-to-failure" or "failure fraction" due to any or all competing PoF mechanisms if occurring simultaneously (see, for example, NBTI, HCI, TDDB and EM on Si DSM technologies). Failure models are characterized by the instantaneous failure rate drafted by the bathtub curve. It is a simple and comprehensive description of how failure distributions are observed during aging.

– Random failure rates are expressed in failures in time (FITs). Mathematically, wearout failure rates tend to increase with time and are characterized by Weibull distributions with shape parameter, $\beta > 1$, while random failure rates are defined to be constant with time, characterized by the exponential distribution (exponential distribution is a special case of the Weibull distribution, in which the Weibull slope is one), or decreasing in time for infant mortality, characterized by a Weibull $\beta < 1$ (VITA 2011).

– Mitigating measures and drift parameter criteria for each failure type may also differ. Wearout can be moderated by limiting exposure to fatigue environments, but this may not help to prevent random failures.

– Both random and wearout failures are described by different Weibull shapes parameters. They need to be combined since these differences limit the usefulness of a single, combined metric. Nevertheless, both metrics must be considered in competition for new emerging COTS devices due to their nanoscale geometries and low voltage biasing range.

– If random phenomena exist, reliability can be estimated using the chi-squared distribution as a function of the cumulative operation and the number of failures observed.

– If wear phenomena exist, it is necessary to adjust the reliability or accelerated degradation model from the operating time hypotheses or the degradation levels observed in tests using the maximum likelihood method.

– It is important to consider if high-level accelerated test conditions that are far removed from the use condition may lead to the activation of different failure mechanisms because their order rank can be altered (see, for example, HCI, activated mainly at low temperature compared to EM which is activated at high temperature).

It becomes necessary to describe a process flow to assess the wearout time-to-failure that a component can achieve before the end of lifetime of the product using it. This tool involves managing and implementing document history (return of experiment, heritage) and storing it to assess the risk a new device will bring to a future piece of equipment in construction and design. This method is based on data and models available about a given technology; thus, it will not give a confident lifetime of a component, only a level of risk that the lifetime is too close to product lifetime.

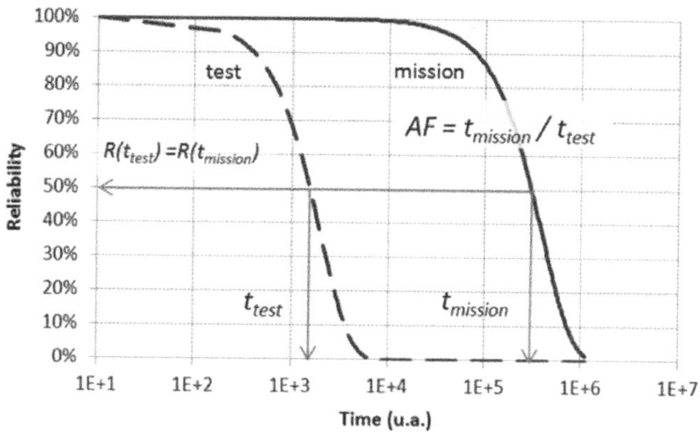

Figure 2.2. *Standard model of accelerated life. For a color version of this figure, see www.iste.co.uk/elhami/mechatronics.zip*

The questions raised in this chapter address some very specific aspects:

How do we build a methodology to forecast the reliability of a device, card or system in harsh operating conditions? We want to forecast and not predict because predicting is something that is related to a "crystal ball".

What are the clues and what evidence is there to correctly and objectively answer if there is a risk or not when operating a new technology, in a repeatable, safe and precise analysis process?

How do we automatize such a method in order to reduce uncertainty and avoid the influence of many expert positions, or even give a single expert the right to return to their analysis years later to improve it using new evidence and data?

2.2.2.1. *Prognostic Failure Methodology (PFM)*

Current predictive reliability standards, such as the 217^+ (QUANTERION 2015) or FIDES (FIDES 2004) quality guideline are based on the Arrhenius and Eyring laws, which express random failure rate and time-to-failure (TTF) as a function of a constant activation energy and power law of stressor parameters. These tools involve accelerated tests that aim to quantify the lifetime-to-failure characteristics of a product, equipment, system, electronic card or component under specific environmental operating conditions, termed "mission profiles". They are intended to predict the constant failure rate (FR) for each failure mechanism.

Confusion may exist as the failure rate of a given mechanism can be associated with two concepts:

– either the end of life or so-called wearout region of the bathtub plot, considering the failure rate is by definition a non-constant failure rate process;

– or the random failure rate period where the devices under mission profile or equivalent accelerated tests may fail randomly, but are still related to a single failure mechanism. Because of the exponential law, the time-to-failure is simply deduced as the inverse failure rate.

The two types of TTF must be considered and quantified. Generally, wearout is supposed to occur later than the TTF deduced from constant FR.

Today, because advanced microprocessor PMOS or NMOS FinFET technologies continue to shrink with very narrow node geometry and process variability close to the operating voltage limits, early wearout has been confirmed in the last decade by Boeing (Condra et al. 2005) and Thales (Regis et al. 2014) (see Figure 2.3).

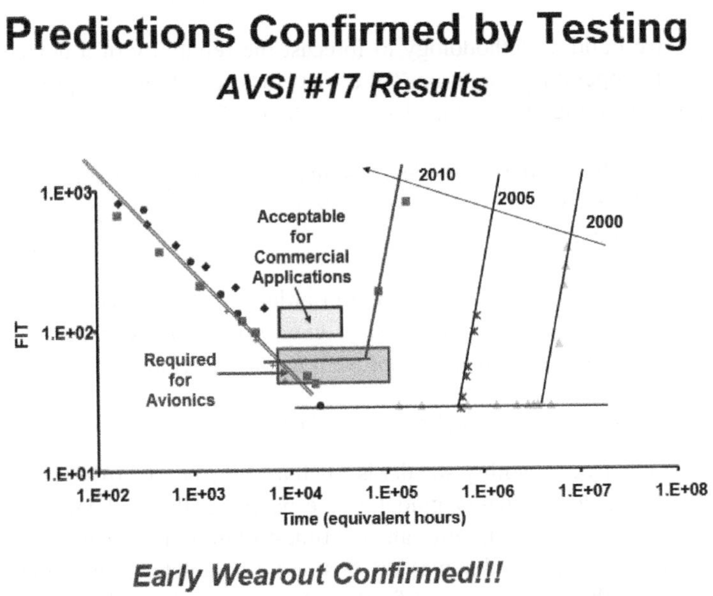

Figure 2.3. *Extract of a presentation at the National Software and Complex Electronic Hardware Standardization conference in 2005 (Condra et al. 2005). For a color version of this figure, see www.iste.co.uk/elhami/mechatronics.zip*

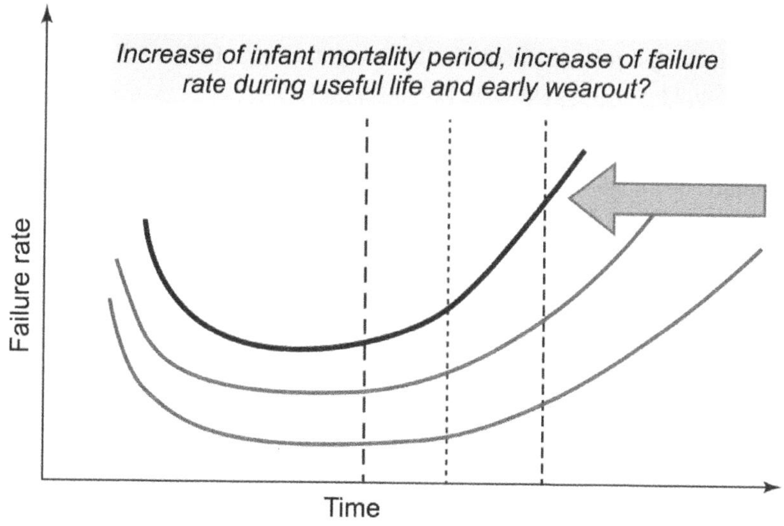

Figure 2.4. *Bathtub plot evolution (Regis et al. 2014)*

Indeed, as noted by Regis et al. (2014), if most publications agree to state that wearout failures occur more and more prematurely and that they significantly reduce useful life, it is more difficult to determine how failure rate behavior will evolve over the next few years. The authors mentioned from all recent studies and publications on Deep Sub-Micron components state at least two facts, for example, on the one hand, an expansion of the early life and, on the other hand, a shortening of the useful life.

Failure rates evolve, as shown in Figure 2.4, which depicts a bathtub curve describing the relative failure rate of an entire population of devices over time.

The lifetime of a component or of an electronic card is quantified using a constant random failure rate (λ). This law of probability applies to rare events or a situation where the manifestation of an event occurs randomly over time with a low probability. The reliability estimation problem turns out to be different if we consider a product subjected to a degradation process that can be quantified over time and on which a criterion of acceptability can be set. It is then possible to track the evolution of this drift during stress tests in order to evaluate the forecast reliability of the product, or during its operational life, to know its state of health from day to day, estimate its remaining potential and act in consequence.

How do we discriminate these concepts (constant vs non-constant rates) when people are locking a pragmatic solution to assess the risk of failure during the entire mission life of a piece of equipment, product or component?

Answering this question requires the development of a failure risk analysis methodology, as initially stated in a guideline proposed by GIFAS (*Groupement des Industries Françaises Aéronautiques et Spatiales*) (W.G. Components 2015). This document describes a method for selecting digital components to ensure that the reliability of these components is compatible with the requirements in aeronautics and aerospace environments. Guidelines are provided to assess the long-term reliability of COTS semiconductor components in such applications. Those guidelines mainly apply during the electronic design phase when selecting electronic components and assessing application reliability. It focuses on the intrinsic wearout of DSM COTS semiconductor components processed below 90 nm and puts aside, at this time, packaging wearout and random failure mechanisms. In this view, physics of failure (PoF) is at the heart of the approach.

To address this demand, a reliability risk analysis of the lifetime of an electronic device versus the mission profile of the final product is defined as a special guideline document named the prognostic failure methodology (PFM).

The general overview method is shown in Figure 2.5. Three loops are drawn from the most general to the most detailed. The external ring starts from the technology under analysis: How is the design at system level defined? What are the constraints and stress environment? Which applications is it for? What are the criteria associated with the failure modes and what are the end-of-life (EOL) prediction or maintenance requirements? Each step is then detailed through data processing sequences and four risk levels are defined, which depend on the reliability and design knowledge that the manufacturer is able to share with the user. The goal is not to calculate the failure rate of a component, but to assess the level of risk of the lifetime being shorter than the mission duration of the COTS electronic component, within its nominal environment.

Reliability concerns can be seen and analyzed according to the level of knowledge and parameter (signature or indicator) measurements:

– from a purely statistical point of view, when a population of devices is described by catastrophic failures counted at random in a stochastic process (no history of the past is taken into account). In this case, constant failure rate λ_c is defined based on a chi² test and the number of hours x the components (e.g. number of devices multiplied by the duration of the observation). So λ_c is expressed in FIT (failures in time or number of failures in 10^9 hrs).

As pointed out by Nikulin et al. (2007) (see Chapter 8), it is natural to consider that a component experiences a wear effect characterized by a growing monotonous failure rate. The contributing link between degradation and particular types of failure can be quantified by two considerations:

– the failure occurs when the degradation exceeds the threshold limit;

– the failure has an instant risk function that depends on the history of component degradation.

The corresponding mathematics that support these two kinds of observation (random or wearout degradation) are quite different, and give distinctive time-to-failure estimations. The constant failure rate λ_c allows us to calculate one equivalent time-to-failure (TTF) as the inverse of λ_c. Another way is to characterize the start of a failure mechanism occurrence by analyzing a series of test experiments under various conditions of stress (assumed representative to accelerate the same failure mechanisms as the one existing under nominal use of a device). In this case, the parameter drift population is described by appropriate statistical models (Weibull, Log–normal, Beta or Gamma functions, etc.). Each model is represented by a quantile value parameter time to reach failure criteria, generally defined as the time for half of the population to fail (the most usual one). This time is noted $t_{50\%}$. Depending on the mathematical model applied, the *MTTF* is calculated according to the following expression:

$$MTTF = E(t) = \int_0^{\infty} t \cdot f(t) \cdot dt = \int_0^{\infty} R(t) \cdot dt \qquad [2.2]$$

where *E(t)* is the mean expected value of the variable t. Refer to Chapter 5 for the definitions of f(t) and R(t). For an exponential distribution:

$$R(t) = e^{-\lambda \cdot t} \qquad [2.3]$$

And equation [2.2] reduces to:

$$MTTF = \int_0^{\infty} e^{-\lambda_c \cdot t} \cdot dt = \frac{1}{\lambda_c} \qquad [2.4]$$

It is also easy to demonstrate that the expression of MTTF for a log–normal distribution ($t_{50\%} = e^{\mu}$, μ is the location parameter and σ is the scale parameter) is given by:

$$MTTF = t_{50\%} \cdot e^{\frac{\sigma^2}{2}} \qquad [2.5]$$

The failure rate for an EOL mechanism is typically a non-constant failure rate that increases with time and then decreases when the lot population reduces to zero; this is because, by definition, the failure rate is the ratio number of failed devices divided by the number of remaining sound devices. This ratio reduces when the remaining population decreases to zero.

Consequently, in parallel to the constant failure rate evaluation, it must be verified that the wearout of a component will occur long after the lifetime of the final product that it is a part of.

Manufacturers may sometimes provide information about the lifetime although, unfortunately, they do not often communicate anything on this. The methodology describes a process flow to assess the wearout time-to-failure that a component can achieve before the end of lifetime of the product using it. This method is based on data and models available about the technology, so it will not give a confident lifetime of the component, only a level of risk that the lifetime is too close to product lifetime.

The method proposes four levels, as defined in GIFAS (2015), depending on the data and service available from the manufacturer. Level 1 is the simplest to implement, and level 4 is the most complex:

Level 1: the original component manufacturer (OCM) provides the assessment of the lifetime of its product using the mission profile that the original equipment manufacturer (OEM) proposes.

Level 2: the OCM gives the means and tools to assess the lifetime, with data or mathematics physical models. Then, the OEM computes and assesses the time-to-failure of the product under the mission environment.

Level 3: the OCM declines to give any information related to data on the reliability models of its products. Then, the OEM must assess the lifetime, based on the component technologies, bibliography, manufacturing analysis and physics of failure models.

Level 4: the product COTS is not mature enough and no reliability data exists, or there are very few which are not consolidated. The OEM must perform an accelerating life test sequence and evaluation program to construct reliability models, identify physics of failure and characterize failure mechanisms.

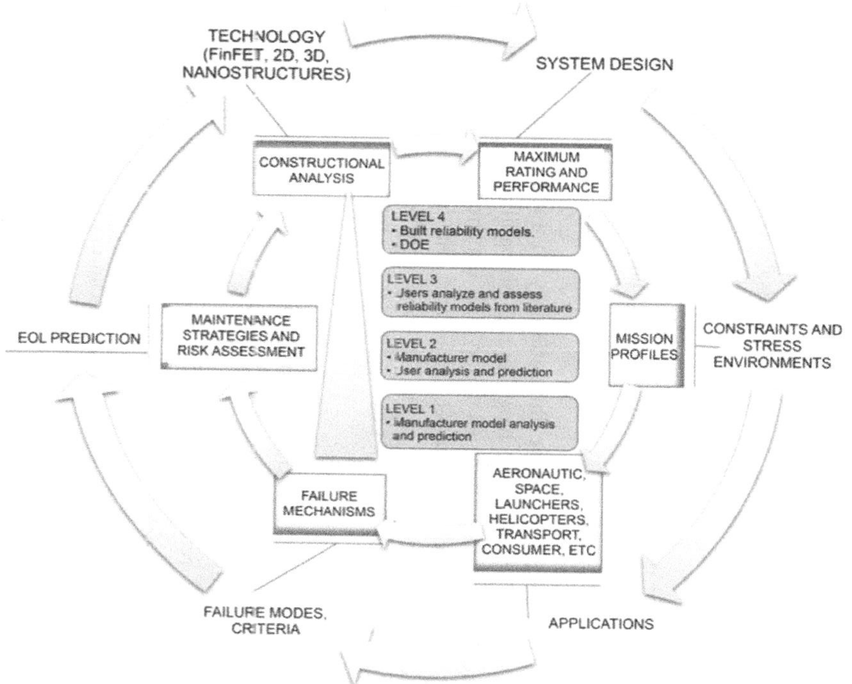

Figure 2.5. *General PFM methodology overview (the external ring is related to the PFM method loop from technology to EOL prediction. The second ring describes the definition/content of each topic. The center relates to four levels of reliability risk assessment). For a color version of this figure, see www.iste.co.uk/elhami/mechatronics.zip*

2.2.3. *Prognostic failure model (PFM) level 3: reliability prediction applied to DSM technologies in harsh environments*

This chapter gives an overview on how to conduct the PFM methodology applied to a specific industrial case.

The goal is to assess the reliability of one component from industry and qualitatively assess the risk to implement such new technologies in aeronautic or space missions for application in 15 to 30 years. Let us take the following example of a component based on advanced 3D technologies, such as three gate CMOS FinFETs (FF) designed at 20 nm nodes and dedicated to SDRAM memory. The proposed study is on:

the DDR3 MICRON, which contains Synchronous DRAM (SDRAM) with LG = Lnom = 20 nm (FF20)

The final objective is to determine the reliability of such a component with respect to wearout mechanisms and hard failure probability in the context of avionics and space applications, that is, under high constraint environments and mission profiles. These harsh constraints arise from the technology parameters themselves, as the components may be placed in boost configurations depending on the supply voltages (multicore processor), operation frequency and bit rate transmission.

The emphasis is on how to validate or improve a technical expertise in the calculations of worst-case degradation modeled under each mechanism, which may be involved in each industrial component and the corresponding lifetime value associated with the most relevant failure with respect to the optimized FinFET technology, and to the use case of memory and FPGA circuits.

The following is a synthetized view on how we can clarify some concepts for a comprehensive harmonization of existing reliability models of failure mechanisms and associated failure mode definitions:

– ***Internal electrical*** **stresses** labeled ***stressor parameters*** are responsible for the wearout failure rate (Weibull $\beta > 1$). There are only four types of applied and imposed stress conditions: they are voltage, current, dissipated power and input signal or ESD/EOS/EMC energies, and can be either static, dynamic, transient or surge. They are quantified with respect to their level of stress applied compared to their level of burnout instantaneous failure mode. But for the sake of standardization and normalization, they are limited by the maximum values allowed by the technology.

– When a device operates under ***external stress*** (thermal management constraints, packaging and assembly constraints, atmosphere contaminants, radiation environments), such stressor parameter levels are modified and can reach their maximum burnout and breakdown limits, thus accelerating wearout failures compared to temperature and biasing stress in the absence of external environment.

– ***Failure modes*** of interest are ***electrical or mechanical signatures*** related to failure mechanisms observed and are ***predictor parameters***. Such parameters can be measured as the absolute drift value of electrical parameters or as a relative percentage of drift (compared to maximum rating limits).

– ***Constant failure rate (random)*** is caused by random defects and random events. The failure rate is modeled by a Weibull shape parameter close to 1, which is equivalent to an exponential distribution law.

– ***Lot-to-lot production variation*** (respectively device-to-device) and performance dispersion from a single manufacturing lot (respectively device) will affect the burnout limits, inducing a change of the percentage of stress applied on a given lot (device), in return. Statistic dispersion will affect the time-to-failure in a similar way (producing the same statistical effect). Such dispersion at lot and device level will impact the remaining useful life (RUL) for some part of the population.

– ***Infant mortality failure*** populations are caused by "defects" and correlate with defect-related yield loss. They are reduced by improved quality manufacturing and by screening.

2.2.3.1. *Stressor definition and normalization*

In a similar way, considering a population of devices submitted to heating will only degrade continuously up to malfunction and failure. But when superposing high (or low) temperature and adequate stressors, the time-to-failure of such a population will reduce. Here, the term "stressors" is defined as the electrical factors applied to the device of concern. Stressors are all limited by technology boundaries defined by the burnout values of each related electrical parameter (breakdown voltage, current overstress and burnout, power burnout, input signal overstress). These stressors can be normalized with respect to their burnout limits and strains are weighted as a percentage of breakdown limits. The main hypotheses, verified by experiments on electronic devices and populations of similar devices, are:

– the physical instantaneous degradation phenomena due to electrical stress above the limits are observed at any temperature and depend on the active zone temperature of the device under test (McPherson et al. 2000a);

– the relative drift of a predictor parameter is a function of time (e.g. square root for diffusion mechanisms) and relates to a failure mechanism activated by temperature and biasing;

– for a biasing set higher and close to the breakdown limit, the two failure mechanisms (e.g. the diffusion and the instantaneous catastrophic ones) are in competition and occur simultaneously; for the sake of simplicity, it is assumed that they are progressively and linearly combined from a pure diffusion mechanism at nominal biasing to a pure burnout at high bias (voltage or current of power dissipation).

2.2.3.2. *Predictor definition*

As mentioned previously, an electrical predictor parameter ξp is defined as the electrical signature (failure mode) of a failure mechanism of interest. Such a parameter can be normalized with respect to its initial value at time zero as a relative drift $\Delta\xi p/\xi p_0$ (in %). Figure 2.6 depicts a schematic drawing showing how predictors are defined and change when stress time elapses. The predictor relative drift shown is an example of actual measurements performed on microwave transistors when submitted to steady-state aging testing (Bensoussan and Suhir 2013). The predictor of each single device is normalized with respect to its initial measurement (mean value) and the failure criterion was 20% drift reached. So, the drawing is set to consider failed devices for all drifts greater than 20%.

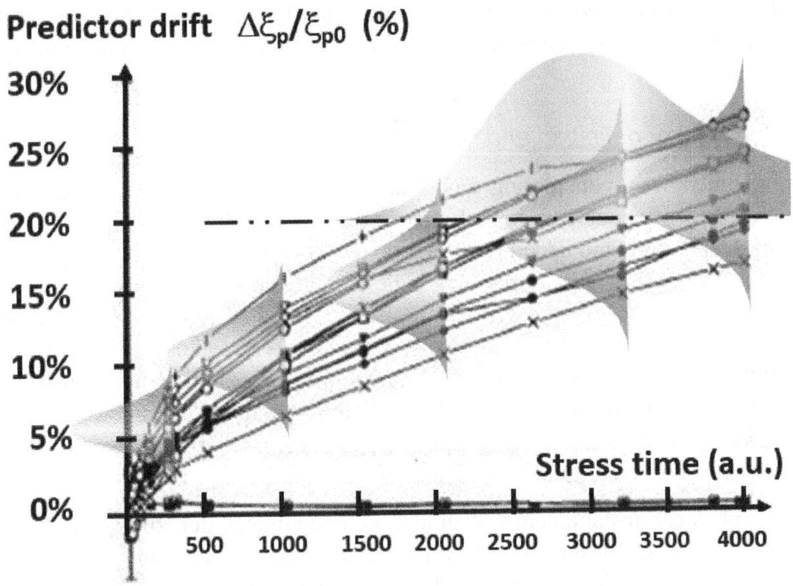

Figure 2.6. *Schematic drawing showing how predictors are defined and vary when stress time elapses*

We define a special industry application in order to:

a) Validate the degradation models firstly identified on the basis of the provided data in the modeling, and de-processing analysis (constructional analysis – CA) for each component. A description of the models adapted to the present cases related to TDDB, HCI, BTI and EM is given, in order to perform a ranking of the wearout mechanisms as a function of the operation mode and in relation to the CA previously performed, which provided details of the structures, dimensions and types of materials.

b) Relate the physics of failure (PoF) to these components, considering their own parameters by giving the necessary details from recent references for justification in correlation to technological data given by the manufacturers. Summaries will be given in tables with respect to selected mechanisms, pointing out the sensitivity and worst-case accelerating values. The goal is to determine a generic structure of elementary FinFET transistors in terms of dimensions and materials as an "intrinsic architecture", which will help us to select the weakest points (at the front-end first) about lifetime calculations according to the operation mode.

c) Understand the data previously published and chosen to develop a data exploitation strategy through justification of the different acceleration factors in AC mode versus DC mode. A validation of the models for avionic and space applications using their respective mission profiles will be given. If possible, the results of the study will be extended by giving new recommendations for design rules carried out by the manufacturers.

Therefore, we first study the validity of the development, focusing on the distinct acceleration degradation modeling, and consider the most adapted framework based on recent developments on the main degradation mechanisms obtained in the literature. We then correlated this part to the FinFET technologies and their own specificities for the use case, which will concentrate on the extraction of sensitive parameters with the help of the accessible data from the manufacturers. Hence, the first part deals with the validity of the chosen accelerated models with respect to published results about wearout and failure mechanisms that are still in progress with the continuous scaling of the CMOS technology.

2.2.3.3. *Component selection*

As the ultimate goal is to select a component with regard to searched functionality and their level of performance, the risk that the lifetime of a component does not comply with the lifetime of a product (regarding its mission profile) becomes highly probable when shrinking the device geometry, and when design rules and biasing conditions are applied. The tested technologies are developed from the recent FinFET 20 nm and 16 nm processing developments, but they have a distinct application scheme. Then, it is necessary to consider the main differences regarding the general application type and extract the main parameters and operation conditions given by the manufacturer that are involved in the wearout damage.

2.2.3.4. *Construction analysis principle with PFM*

The principle of dividing the PFM methodology into sub-sections between levels 1 and 4 is relevant considering the low level of knowledge we may have from the manufacturers. The difficulty is indeed about the correlation to the limited results published on new FinFET structures with the theoretical acceleration laws, which may be applied to these 3D devices, pointing out the good correspondences and limitations of the techniques. This is also linked to the type of devices (N-channel vs. P-channel), the shrink of the channel length (L_{eff}/L_{nom}) and the processing flow (dummy gate, replacement metal gate, trench isolations for the gate and for lateral isolations) that may continuously change with respect to the trade-off between performance and reliability (Randomson et al. 2007) done by the manufacturer.

Many aspects are addressed in detail in the whole PFM methodology and the principle of lifetime calculation considering a realistic mission profile that accounts for the on/off phases, standby phase and aircraft maintenance. For example, to express the TDDB, the known reference acceleration factor, the temperature, electric field and device geometry acceleration factors can be determined separately. This is a rule of thumb that will give us the technology quality and resistance under a given degradation mechanism, considering the effective temperature. However, we have too many unknown parameters from the technology side, which confuse the determination of an accurate lifetime value for a given device and functionality. That is why a sort of shortcut is implemented in order to simplify the lifetime extractions for all mechanisms that need to be selected in order of importance for memory application (DDR3L).

Consequently, we propose lifetime extractions based first on the selection of the mandatory elements required for an accurate analysis of reliability by predictive calculations applied to these recent technologies, that is, based on 3D FinFETs 20 nm down to 16 nm. This can be categorized by numerical calculations of a minimum, typical and maximum reliability parameter extraction in states of dimensions, voltage, temperature that will be related in a first step to uncertainties arising from processing quality and technological spread in designing FF at such dimensions. Indeed, this architecture is often obtained by a simple shrink from previous CMOS nodes, as we do not know the full process flow. For example, in some CMOS products, TSMC has derived its 16 nm FinFET (FF16) CMOS from the shrink of the previous 20 nm node for Xilinx. That is why the presented results have the finality to be compared to the measurement steps (see Figure 2.7) within a simplified building block framework (similar to level 4) in order to validate the PFM approach, which is becoming strongly complex with the large number of parameters and failure mechanisms (FEOL/BEOL), with the range of uncertainties extracted from the CA step. This explains why we can propose a three-loop methodology that can be derived from PFM flow, where for step 1 in Figure 2.7, these dimensional uncertainties can be directly transferred to lifetime margins once we know the nominal supply voltage (V_{DD}) and boost conditions in speed (V_{DDmax}) between the core and IOs.

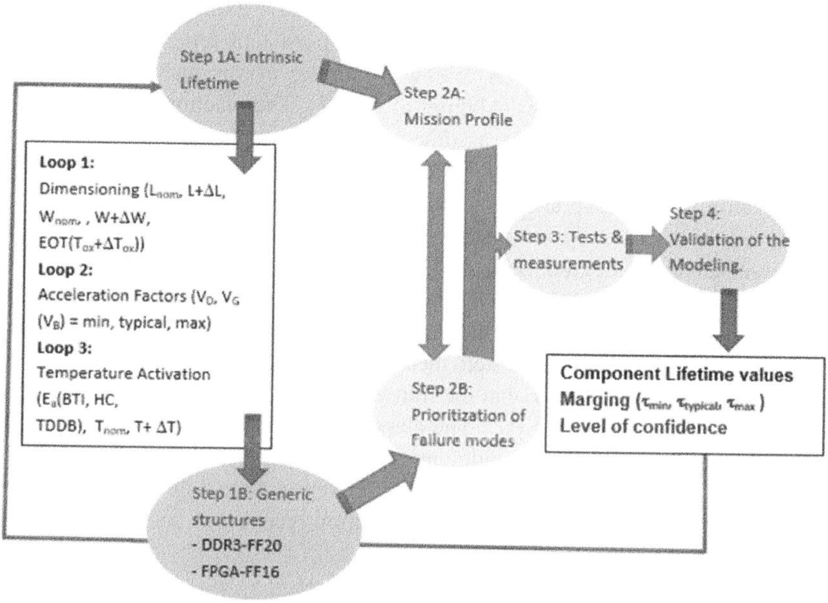

Figure 2.7. *Schematics of the main steps involved in the reliability determination of DDR3 (FF20) and FPGA (FF16) CMOS technology nodes. For a color version of this figure, see www.iste.co.uk/elhami/mechatronics.zip*

Hence, this study is primarily devoted to the realization of step 1 and step 2, according to the tasks cited above (work plan) and the possible comparison later to measurements performed (step 3) under realistic operational temperature and cycle conditions. It is important to see that the PFM approach and lifetime determination for components require this comparison step, which can bring good calculation accuracy through temperature dependence and supply voltage acceleration.

2.2.3.5. *Degradation model monitoring*

As applications such as aerospace, vehicles, implantable medical devices require zero failure, the reliability assurance of the future, which must be based on the transitioning from time-to-failure modeling to degradation modeling, must be taken into account. The Gibbs free energy model presented in Chapter 1 is a method for reaching a lower potential and describing degradation physics. Because of thermodynamic laws, the first law of energy conservation and the second law of entropy as a description of the disorder of isolated systems tends to increase with time. The Gibbs potential report changes in enthalpy and entropy, and so degradation will always lower the Gibbs potential because all forces on electrons/atoms/molecules are derived from gradients (McPherson 2019). The degradation models commonly used are power law or exponential or logarithmic laws. As for NBTI, this mechanism has been a persistent reliability concern for silicon ICs since the mid-1960s. A review of studies of various aspects of NBTI degradation has been reported by Alam et al. (2007). The reaction–diffusion (R–D) formulation of NBTI degradation assumes that NBTI arises due to hole-assisted breaking of Si–H bonds at the Si/SiO_2. Alam et al. address a careful analysis of the absolute value of time-exponent *n*, and determined that the diffusion of the H_2 molecule would consistently interpret experimental data regarding quasi-saturation, frequency-insensitive degradation and $n = 1/6$ exponent with "no-delay" measurements. Yet, they also considered that "instead of absolute values of *n*, the groups of studies who focused on the variation of *n* as a function of temperature (determined by standard delay-based measurement) reached an entirely different conclusion (Kaczer et al. 2005): that the transport of H-specie is dispersive and most likely specie is H+(proton), not H_2". The classic exponent of $n = 1/4$ is the signature of NBTI degradation based on a wide variety of experimental results. But it is well known that N_{IT} (the fraction of Si–H bonds broken at time *t* due to NBTI stress) generated during the stress phase of the degradation recovers as soon as the stress is removed: NBTI is a self-annealing process.

Considering interface trap generation with stress time for different models of hydrogen diffusion, the RD model summarizing interface trap generation in all of the regimes of stress phase is shown Figure 2.8 (from Alam et al. (2007)) and also plotted the various cases of resulting power laws. The exact value of the slope depends on the underlying mechanism of hydrogen flow inside oxide and the mechanism of reaction.

Stathis et al. (2018) concluded that different approaches to the fundamental study of NBTI naturally lead to different conclusions:

> There is still no consensus on the basic physics of negative bias temperature instability. Two competing models, reaction–diffusion and defect-centric, currently vie for dominance. The differences appear fundamental: one model holds that NBTI is a diffusion-limited process and the other holds that it is reaction-limited. Basic issues of disagreement were summarized, and the main controversial aspects of each model were reviewed and contrasted.

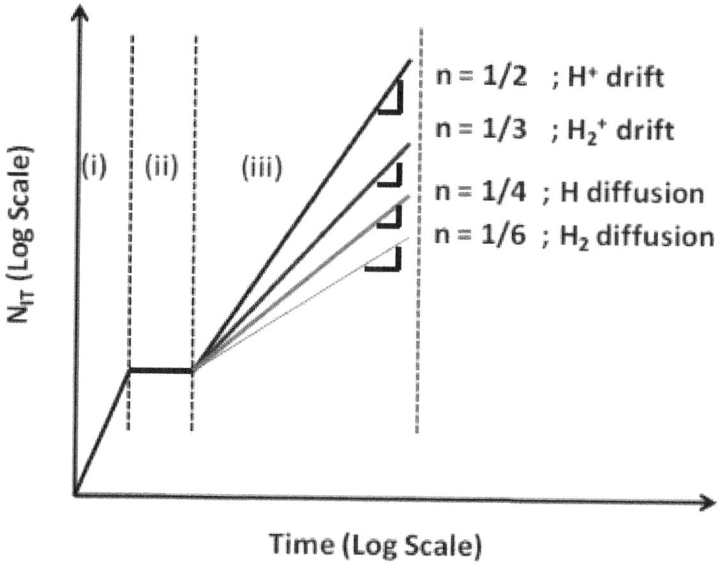

Figure 2.8. *Interface trap generation with stress time for different models of hydrogen diffusion. For a color version of this figure, see www.iste.co.uk/elhami/mechatronics.zip*

Broadly speaking, the critiques of these two models can be summarized thusly:

(1) While the reaction–diffusion (RD) model can now very successfully describe a large variety of observed data over a broad set of experimental conditions, the validity of the underlying physical interpretation is questioned because the model parameters conflict with the established literature on H in Si/SiO$_2$ systems.

(2) The defect-centric (DC) model has a strong basis in physics, supported by the microscopic measurement of discrete defects, but until recently, it has paid scant attention to interface state generation, and the ability to comprehensively describe NBTI stress data over a broad set of process and stress conditions is questioned.

There is a need to be selective with respect to all degradation mechanisms (assumed here to be described by the constant failure rate mechanisms) that can be involved at the frontend and backend. Generally, FEOL reliability requires mainly TDDB, BTI, HC and EM at the first level of metallization (M1), but may also consider cosmic rays (CR) through the gate reaching active channel and S/D regions. BEOL reliability analysis generally focuses on interconnect stacks looking at electromigration (EM), stress migration/voiding (SM/SV), thermal mechanical stability, low K dielectric breakdown (TDDB) and chip/package interactions (CPI), while cross talk and electromagnetic compatibility (EMC) can also have a net impact for a given packaging.

In this present work, we focus on FEOL. Until now, only a few works deal with interactions between degradation mechanisms, considering that each progressive degradation can be treated separately, leading to a direct simplification that the wearout issue of a system (MTTF) examined by a steady-state failure rate (λ_{SS}) can be considered as a generic failure rate λ_G for device "*i*" expressed as a function of its own acceleration factors (*AF*) depending on voltage, temperature, quality (variability) and active surface with:

$$\lambda_{SSi} = \lambda_G . AF_V . AF_T . AF_Q . AF_S \quad [2.6]$$

leading to the sum of a total failure rate $MTTF = 1/\lambda_{SS}$, with:

$$\lambda_{SS} = \pi_E \sum_{i=1}^{N} n_i \lambda_{SSi} \quad [2.7]$$

where n_i is the quantity of device "*i*", *N* is the total number of devices and π_E is the environmental factors that can be related to mission profile (t_{on}/t_{off}). Our goal is to determine the component reliability regarding its generic reliability criterion based on a (simplified) generic structure extracted from the DDR3 and FPGA components.

2.2.3.5.1. TDDB modeling

Time-dependent dielectric breakdown (TDDB) in dielectrics is one of the more important failure mechanisms for digital circuits, as the scaling of metal-oxide-silicon field effect transistors (MOSFETs) is becoming a figure of merit for performance gain with miniaturization. For this reason, physics-based models for TDDB and physics of failure (PoF) include many different approaches through gate-oxide thinning with field-based models (McPherson et al. 2000b), current-based models (Lee et al. 2005) or a combination of field and current-based models (Huard et al. 2005), as well as a voltage power law acceleration model (Nicollian et al. 1999; Nicollian 2007) and multi-vibration H release (MVHR) model (Ribes et al. 2006).

The PFM development for the TDDB study at FEOL is well organized, taking into account the most important parameters that are primarily involved in lifetime determination and pointing out the difference between use conditions and accelerated stress conditions. Good references are chosen throughout the historical TDDB modeling, giving a clear picture of the recent trends used for lifetime extrapolation. For practical use of the TDDB calculation, the way by which the slope Weibull plot (β) is obtained, as well as the level of confidence at $t_{63\%} = \eta$ factor extracted from publication can bring a certain level of accuracy depending on the publications used. The difficulty arises from the lack of recent publications on FinFET TDDB coming from foundries such as TSMC, Intel and Samsung. Even if the last published models are not referred to as, for example, clustering (Wu et al. 2013) and filamentary modeling (Blonkowski 2010), this leads to bi-modal distributions (β variation with T_{ox}). This recent modeling brings more accuracy for a high level of percentile, non-Poisson area scaling. This is explained by the non-uniform distribution of defects (deviating from Weibull statistics) between the interfacial layer mode and bulk oxide behavior that result in a change of slope and related level of confidence through η factor variation, which leads to saturating behavior in AF_V influence. As we want accuracy towards small supply voltage conditions, we can indeed remain on the E model for our subject of interests, especially for TDDB.

One focus is to determine the lifetime margin (TDDB) with typical, min and max values (that can be related to voltage conditions between derating and boost). We can indeed (as proposed by PFM) choose the most severe lifetime criterion as the E-model (thermochemical), which is most restrictive with respect to its direct acceleration with oxide field (exponential law), and is satisfactorily applied to low field and long term. The 1/E model is rather straightly related to the injected (gate tunneling) current at high field, and was historically used in thicker gate-oxide MOSFET (T_{ox} > 5nm), leading to a direct determination of t_{BD} in relation to the anode hole injection (AHI) mechanism that puts the back-injected holes into the modeling (NMOS device) (see Figure 2.9).

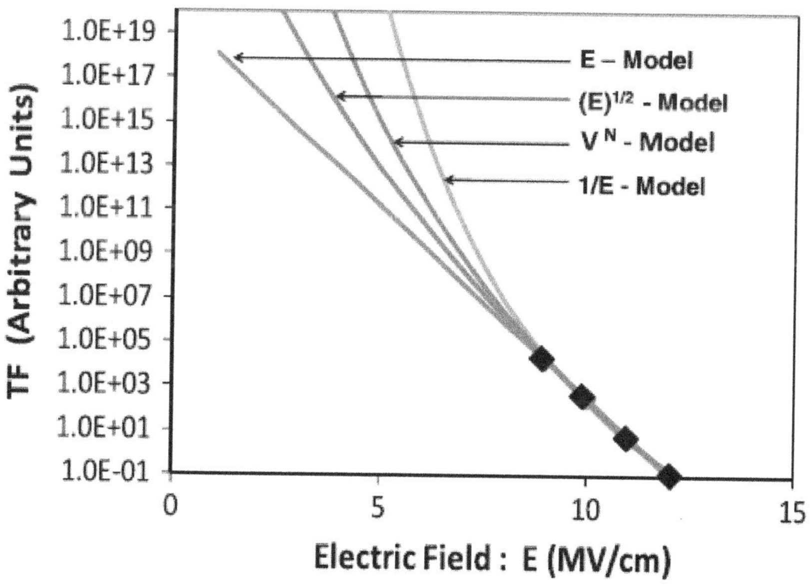

Figure 2.9. *Comparison of different TDDB acceleration modeling (McPherson 2012). For a color version of this figure, see www.iste.co.uk/elhami/mechatronics.zip*

However, with T_{ox} thinning (< 3.5 nm), the charge at breakdown (Q_{BD}) is strongly reduced due to direct tunneling leakage currents. This is why some works turned out as a direct voltage dependence, either as an exponential law (Monsieur et al. 2002; Ribes et al. 2006), or a power law in gate-voltage (Wu and Sune 2005), known as anode H release modeling. Therefore, we indeed prefer the E model as it remains more restrictive for lifetime estimation in products, especially devoted to avionics and space.

Moreover, with the reduction of supply voltage, many results have shown that hard breakdown (HBD) from the gate-oxide field has a reduced probability to occur, and that when it does occur (coming also from a triggering of field spikes or the correlation to other failure mechanisms such as ESD, EM at first metallization level M1), it would first induce a soft breakdown (SBD) that can reach HBD, but would still maintain functionality, particularly in ring oscillator or digital cells, showing that the first breakdown event does not necessarily spell the immediate failure of the entire CMOS application (Kaczer et al. 2007).

This points out that HBD may be more closely related to BEOL reliability and the scaling challenges of thinning ILD for metallization with the use of the low-K that has to be examined as a function of the type of material (SiOCH) and its level of porosity, the mechanical weaknesses such as cracking, delamination that lowers the TDDB breakdown strength, and also the critical spacing rules for metallization M_x to M_{x+1} level. All of these mechanisms with scaling can have a stronger effect for failure analysis due to resistance variation with material quality (grain boundary, surface scattering and increase of liner/Cu volume ratio).

This is a reason why HBD for FEOL and core using low supply voltage is not relevant when applied directly to the device from M_0, but rather applied to IO with larger supply voltage. In a way, the more we have on current, the more we meet the dissipation process and hot spots that link us to FEOL reliability issues at the M1 level with ILD and low-K that suffers from smaller critical field conditions to eventually trigger HBD to gate-drain the overlap region of a biased device.

An example can be obtained with the E-model in 22 nm FinFET, which can be used for DDR3 comparison following the exponential law in Figure 2.10 (data given from INTEL (Ramey et al. 2013)).

In this case, we can directly extract the exponential law parameter (equivalent to the E-model) usually expressed as $exp(-\gamma E_{ox})$ if we want to retain the restrictive evaluation criterion. The practical use is to plot the results expressed as a function of the stress gate voltage VGS (as $E_{ox} = (V_G - \phi_{ms} - 2\phi_p)/E_{OT} \cong (V_G - V_{TH}+0.2)/E_{OT}$), which shows that the TTF_{BD} at low field is obtained as:

$$TTF_{BD} = A_i \cdot e^{-\gamma \cdot V_G} \cdot e^{-\frac{E_a}{kT}} \cdot e^{[Ln(-Ln(1-F))/\beta]} \quad [2.8]$$

NMOS	PMOS
$\gamma = 34.5$ V-1	$\gamma = 32.8$ V-1
$E_{a,N} = 0.6$ eV	$E_{a,P} = 0.6$ eV

Where the last exponential factor is to mitigate the time to breakdown value by the distribution function. We also find the power law in V_G that can be useful, but for different technologies from FinFETs, such as in bulk Si and standard CMOS components (Wu and Sune 2005; Saliva et al. 2015).

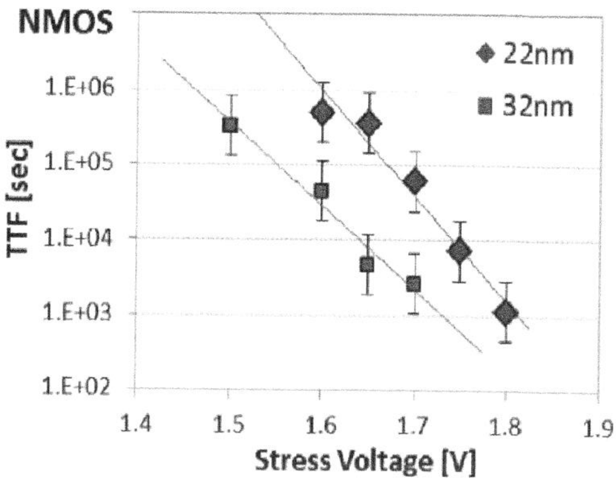

a) NMOS

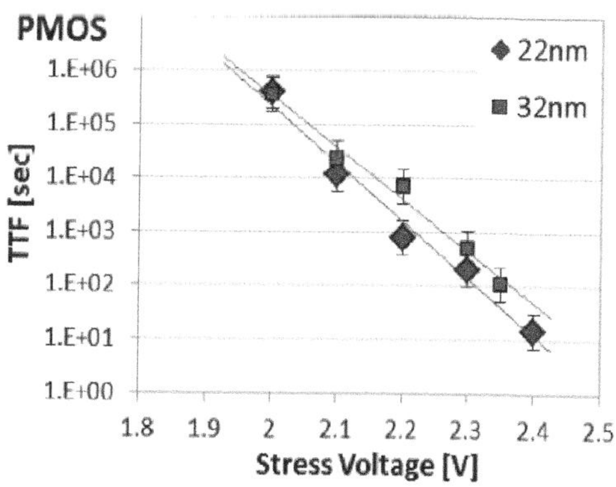

b) PMOS

Figure 2.10. *TDDB in 22 nm FinFET shown as better improved with respect to 32 nm that may be used for DDR3 (Ramey et al. 2013). For a color version of this figure, see www.iste.co.uk/elhami/mechatronics.zip*

2.2.3.5.2. HC modeling

HC modeling may be applied in an adapted way to FinFETs depending on the supply voltage range (V_{DDmin}, V_{DDmax}) and the chosen accelerated conditions (10–35% from V_{DD}) that can be observed from the last references on this topic (Bravaix et al. 2015), as devices operate at low voltage operation:

– V_{DDmax}= 1.5V down, V_{DDmin}=1.28V for DDR3L core memory;

– V_{DDmax}= 0.9V down, V_{DDmin}=0.75V for FPGA core memory;

– V_{DDmax}= 2V for output drivers for HP IO banks of the FPGA.

As we depend on the accessible data of FF reliability for 16 nm and 20 nm CMOS nodes, we can choose the voltage dependence, drain current dependence or both laws including the gate-length acceleration factor, when the data are available.

For the voltage condition of periphery NMOS devices with longer gate-length (Leff = 0.1 µm), common HC determination can be done as this corresponds to the planar CMOS bulk with thin SiO_2 gate-oxide (1.5nm), neglecting the possible use of SiON in this stack. Then, later, we may consider the standard HC lifetime by using a Takeda plot (Takeda and Suzuki 1983), where this useful modeling links the dominant (permanent) degradation mechanism (here ΔN_{IT}) to the maximum lateral field (E_{Max}) with the measurement of impact ionization (*II*) rate, which is directly proportional to (I_{SUB}/I_{DS}). This is generally done as a function of the stress condition (which is maximum in the AC transient during operation for the DC/AC transfer function) corresponding to the maximum of HC rate ($V_{GS} \approx V_{DS}/2$). It results that the lifetime expressed as τ - (I_{SUB}/W_G) is equivalent to τ - $exp(A/(V_{DS} - V_{DSat}))$, where the most common approximation gives the simplified form as τ - $exp(B/V_{DS})$, if we do not know the exact value of V_{TH} is saturation (including DIBL effects (Singh and Samanta 2012)). This modeling is referred to as the lucky electron model (LEM), where if we have the data of I_{SUB} and I_{DS} we have the classical relationship:

$$\frac{\tau \cdot I_{DS}}{W_G} = C_1 \cdot \left(\frac{I_{SUB}}{I_{DS}}\right)^{-m_1} \quad \Leftrightarrow \quad \frac{1}{\tau_{M_1}} = C_1' \cdot \left(\frac{I_{DS}}{W_G}\right) \cdot \left(\frac{I_{SUB}}{I_{DS}}\right)^{-m_1} \qquad [2.9]$$

We call this Mode 1 (M_1) for the "high voltage" condition, that is, $V_{DD} \geq 2.5$V (L_G = 0.25 µm). For voltage with $T_{ox} \geq$ 5nm, there is thick gate-oxide where charge trapping occurs. In medium T_{ox} thickness, high-field-assisted charge de-trapping modifies trapping efficiency and kinetics, while in thin-to-ultrathin insulators, T_{ox} < 3.2 nm tunneling effects largely increase, modifying the classical HC scenario because of slow traps. We must use the M_2 modeling, taking into account the electron–electron scattering (EES) (Rauch and La Rosa 2005; Bravaix et al. 2010) for medium voltages, while in our case, we consider the low voltage modeling related to the multi-vibrational excitation of SiH bonds at the interface of the dielectric and silicon, as shown in Figure 2.11 in Si bulk.

$$\frac{\tau \cdot I_{DS}^2}{W_G} = C_2 \cdot \left(\frac{I_{SUB}}{I_{DS}}\right)^{-m_2} \Leftrightarrow \frac{1}{\tau_{M_2}} = C_2' \cdot \left(\frac{I_{DS}}{W_G}\right)^2 \cdot \left(\frac{I_{SUB}}{I_{DS}}\right)^{-m_2} \quad [2.10]$$

$$\frac{1}{\tau_{M_3}} = C_3' \cdot \left[(V_{DS} - \hbar\omega)^{0.5} \cdot \left(\frac{I_{DS}}{W}\right)\right]^{\frac{E_B}{\hbar\omega}} \cdot e^{\left(-\frac{E_{emi}}{kT}\right)} = C_3' \cdot \left[(V_{DS})^{\frac{a_3}{2}} \cdot \left(\frac{I_{DS}}{W}\right)\right]^{a_3} \cdot e^{\left(-\frac{E_{emi}}{kT}\right)} \quad [2.11]$$

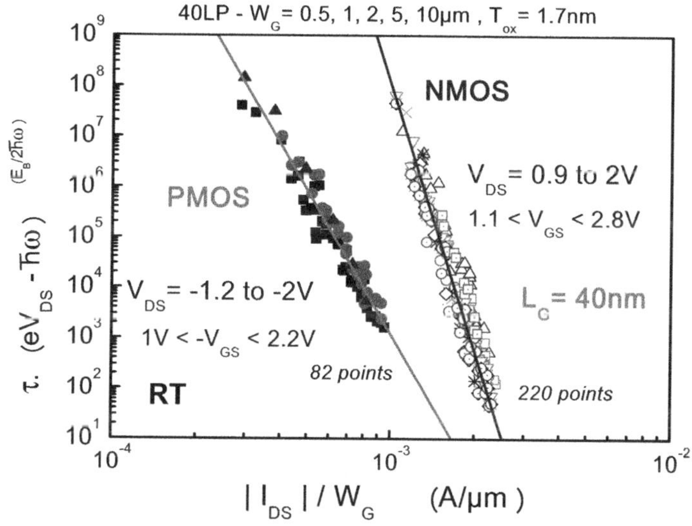

Figure 2.11. *Mode M_3 in C_{40} under CC damage according to [2.11] for a large set of data in PMOS and NMOS. The M_3 degradation dependence gives a_3 = 9.4 and 17.4, with C_3, V_D = $7.5 \cdot 10^{-26}$ and $6 \cdot 10^{-45}$ respectively. The same lifetime criterion 10% ΔI_{DSat} FWD and various stressing V_{GD} can be used for the long-channel Peri COS standard from transistors in inter blocks (Figure from Bravaix et al. (2010)). For a color version of this figure, see www.iste.co.uk/elhami/mechatronics.zip*

To summarize the different parameter extraction for NMOS and PMOS under mode M3 with relations II.8–II.11, we have put the needed values of the degradation modeling in Table 2.1 for the complete lifetime modeling based on the degradation rate between modes M_1 and M_3.

Degradation Modes	NMOS	PMOS
M ($M_1 = M_2$)	2.7	1.3
$\mathbf{a_1}$	1	1
$\mathbf{a_2}$	2.5	3
$\mathbf{a_3}$	17 - 20	9 - 12
$\mathbf{E_{emi}}$**(eV)**	0.22-0.26	0.22-0.26

Table 2.1. *Summary of the parameter in the full lifetime modeling in silicon bulk planar technologies as a function of VDD range and thick-to-medium and ultra-thin gate-oxides (Bravaix et al. 2015)*

We have to point out that a recent publication of HC damage in FinFETs (Chang et al. 2019) has shown that our low voltage modeling (M3) is clearly followed for 14 nm n-FinFETs, with a very similar process as the HfO_2/TiN gate stack, where n-FinFETs were fabricated by 14 nm technology with a gate-last process with thermal (IL) oxide (SiO_2) with a thickness of 1 nm. The TiN layers were deposited by ALD as work function metal (WFM) layers. The sizes of the FinFET used were W = 154 nm and L_G = 60 nm and 14 nm. The authors found a -23.4 slope (see Figure 2.13) for L_G = 14 nm (W_G = 154 nm) according to the low energy range for NMOS device lifetime in mode 3, where MVE is dominant.

This graph (see Figure 2.13d) (Chang et al. 2019) will be used for low voltage HC for FF 16 nm, as well as for FinFET 20 nm, as the stress voltage has been used down to $V_G = V_{DD}$ = 1.5V, (DD3R) and V_{DD} = 0.9V to 0.85V in FPGA as a function of I_{On}, enabling the constant extraction Ci process required for lifetime modeling and extraction.

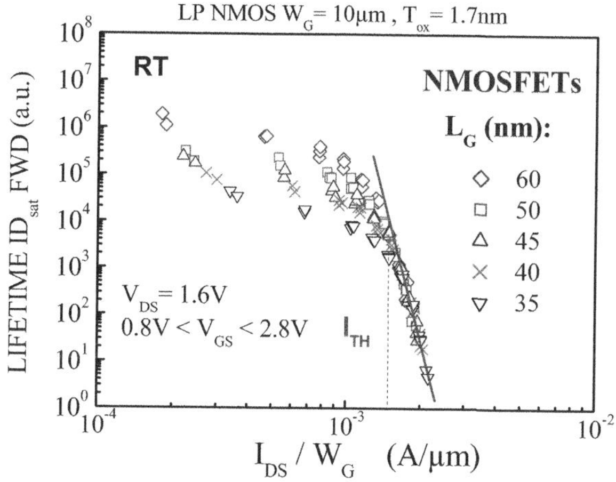

a) When the channel length is reduced

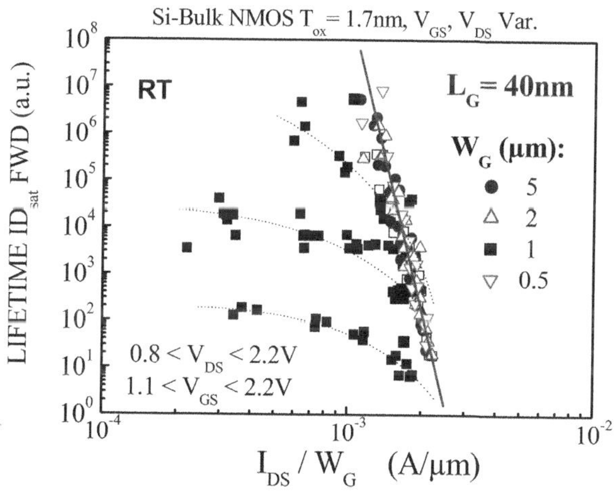

b) When the channel width is reduced

Figure 2.12. *Low voltage HC lifetime as a function of the drive current. This can offer the basics of comparison for standard CMOS (Peri CMOS) of the DDR3L component. For a color version of this figure, see www.iste.co.uk/elhami/mechatronics.zip*

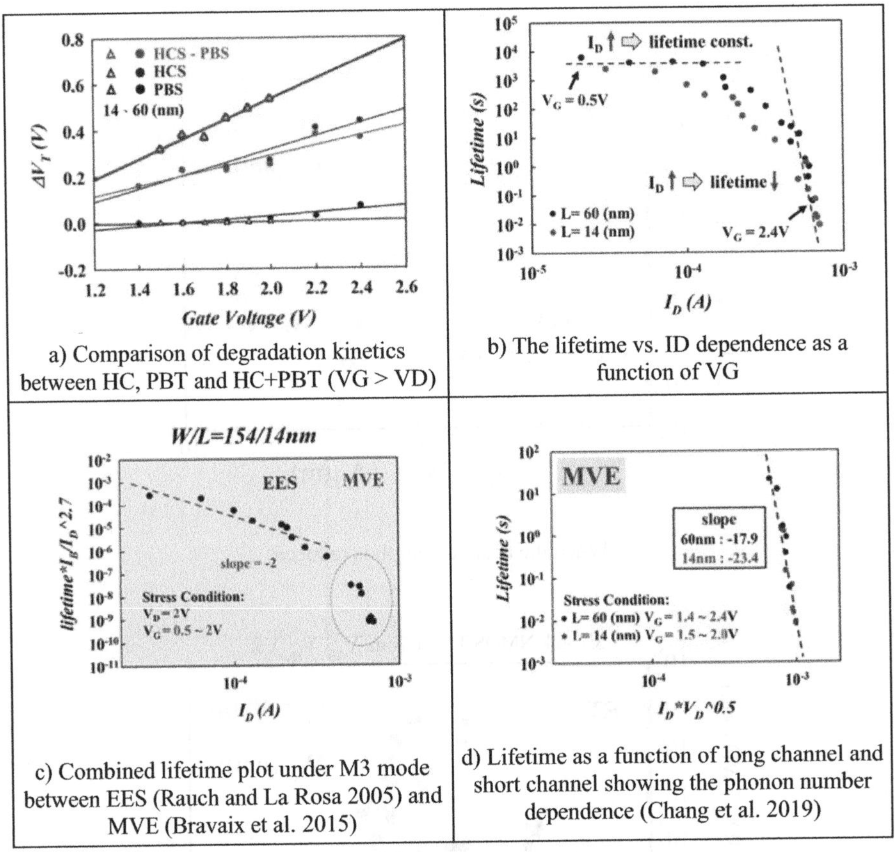

a) Comparison of degradation kinetics between HC, PBT and HC+PBT (VG > VD)

b) The lifetime vs. ID dependence as a function of VG

c) Combined lifetime plot under M3 mode between EES (Rauch and La Rosa 2005) and MVE (Bravaix et al. 2015)

d) Lifetime as a function of long channel and short channel showing the phonon number dependence (Chang et al. 2019)

Figure 2.13. *HC damage in 14 nm n-FinFETs (Chang et al. 2019). For a color version of this figure, see www.iste.co.uk/elhami/mechatronics.zip*

2.2.3.5.3. BTI modeling

BTI degradation is analyzed in PMOS under negative gate-voltage (NBT) most of the time, as this represents the worst issues in CMOS technologies compared to smaller damage under positive gate-voltage PBT.

Among several modified approaches, three models for BTI have historically been developed through technology scaling:

– the reaction limited model with dispersive bond-breaking energies (Huard et al. 2005) and uncoupled (hole) trapping that is [N]-dependent from interface traps (ΔN_{IT}) (Huard 2010); this later being sensitive to H2/D2 anneals;

– the composite reaction–diffusion (RD) modeling (Alam 2009; Mahapatra and Rashmi 2018);

– the stochastic reaction–diffusion (RD) model (Bravaix et al. 2015) improved with a two-stage model and triple well transitions (Grasser 2015) with volatile defects (Grasser et al. 2013).

The general way is to simplify the approach by a level of damage for the threshold voltage related to a lifetime defined at ΔV_{Th} = 50mV with the permanent and recoverable contributions. This corresponds to the static degradation mechanism under NBT accelerated by the vertical field (no channel carriers), meaning that NBT damage is involved with a permanent part and a recoverable part $\Delta V_{Th}\ (t_s) = \Delta V_{Th,\ P} + \Delta V_{Th,\ R}$. If we consider the Huard modeling for total damage, we have the permanent (ΔN_{IT}) and recoverable parts (ΔN_{ot} hole trapping and slow traps) as a function of time constant for capture and emission, stress and recovery time:

$$\Delta D_T = \Delta D_p + \Delta D_R \qquad [2.12]$$

$$\Delta D_P = K_p \cdot V_G^a \cdot t_S^b \cdot e^{-\frac{E_{a,P}}{kT}} \qquad [2.13]$$

$$\Delta D_R = K_R \cdot V_G^c \cdot Ln\left(1 + \frac{t_S \cdot \tau_e}{t_R\ \tau_c}\right) \cdot e^{-\frac{E_{a,R}}{kT}} \qquad [2.14]$$

Grasser has a full (atomistic) description, but a relatively similar dependence for the recoverable part expressed as (Grasser et al. 2013; Grasser 2015):

$\Delta V_{Th}\ (t_s) = P + R$ with $R = R_o/(1+B.(t_r / t_s)^{\alpha})$.

Let us simplify the approach to the main role of the permanent damage that is involved in the long term, in order to determine the NBT impact by the use of the compact model. This can first be considered as a power law for TTF $\propto$ A.V_G^{-m} or else similarly to the TDDB as following an exponential law TTF $\propto$ exp(-B'.E_{ox}) $\propto$ exp(-B.V_G), with the usual thermal factor exp(-E_a/kT). If we consider the shift as the power law in ΔV_{TH} = A t_s^n (n= 1/4 for H°, n= 1/6 for H_2, n= 1/5 as a typical value retained for a mix of both contributions), the pre-factor A = ΔV_{TH}. t_s^{-n} can be normalized to the reached criterion ΔV_{TH}= 0.05V (or ΔID_{Sat} = 10%), leading at 125°C to:

$$TTF_{NBT} = A_i \cdot \left(\frac{\Delta}{0.05}\right)^{1/n} \cdot e^{-B \cdot V_G} \cdot e^{-\frac{E_a}{kT}} \qquad [2.15]$$

with:

Core	IO
$n = 0.16$ (core $\Delta V_{TH} = 50mV$)	$n = 0.15$ ($\Delta ID_{Sat} = 10\%$)
$A_i = 1.48\ 10^8$, B = 12	$A_i = 5.02\ 10^7$, B = 4.6

With apparent activations:

$E_a = 0.35$ to 0.25 eV (Si bulk);

$E_a = 0.6$ eV (FDSOI);

$E_a = 0.15$ eV (FinFETs).

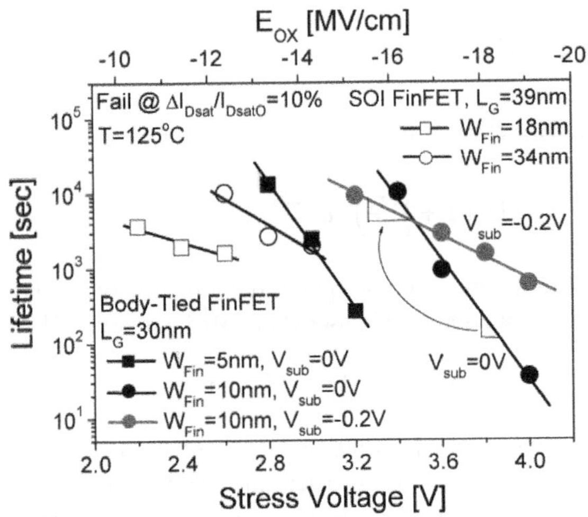

Figure 2.14. *Device lifetime versus stress voltages and E_{ox} of 18 and 34 nm fin-width SOI FinFETs (blank symbols), as well as 5- and 10-nm fin-width body-tied FinFETs (filled symbols) at 125°C substrate temperature. The gray symbol represents the body-tied FinFET with a 0.2-V substrate bias. With minus substrate bias, the quasi-floated body effect is shown (Lee et al. 2005)*

Finally, many types of FinFET structures exist and render the parameters very strongly technology-dependent. The results are shown in Lee et al. (2005). Figure 2.14 shows the device lifetime as a function of stress voltages and of 18 and 34 nm fin-width SOI FinFETs, as well as 5 and 10 nm fin-width body-tied FinFETs at a 125°C substrate temperature. Both SOI and body-tied FinFETs show that the lifetime is more seriously degraded with a narrower fin. The slope of the predicted body-tied FinFET lifetime is steeper than that of the SOI FinFET. This implies that the body-tied FinFET used in DDR3 is more robust to NBT stress than the SOI FinFET. A body-tied FinFET shows a smaller NBTI degradation that becomes more effective due to the presence of the grounded substrate.

With the principle of quasi-static (QS) transfer (Bravaix et al. 2010), inverter waveform can be used in Figure 2.15 to include the off-mode phases, as it represents a general form of digital operation. With the QS scheme, the results of previous sections have evidenced the high V_{DG} conditions required to observe the off-mode degradation in contrast to CHC damage, which needs smaller V_{DS} and much reduced V_{GD} in MOSFETs (Guerin et al. 2007; Bravaix 2019a). Hence, the off-mode damage can be included in the full modeling for digital operation. This is performed by completing the three-mode on-state lifetime (τ_{ON}) developed here for NMOS and PMOS by the off-state contributions (τ_{OFF}) in two phases $1/\tau_{Off} = 1/\tau_{VG=0} + 1/\tau_{SubVT}$ useful for any pulse waveform shapes. Device lifetime extraction (τ) is then deduced for all stress conditions by assuming that all degradation modes compete in parallel, leading to a general device lifetime modeling expressed by:

$$R_{it} = \frac{1}{\tau} = \sum_i \frac{1}{\tau_{on,i}} + \sum_j \frac{1}{\tau_{OFF,j}} \quad [2.16]$$

$$\frac{1}{\tau_{ON}} = C_1 \cdot \left(\frac{I_{ds}}{W}\right)^{a_1} \cdot \left(\frac{I_{bs}}{I_{ds}}\right)^{m_1} + C_2 \cdot \left(\frac{I_{ds}}{W}\right)^{a_2} \cdot \left(\frac{I_{bs}}{I_{ds}}\right)^{m_2} + C_3 \cdot V_{DS}^{a_3/2} \cdot \left(\frac{I_{ds}}{W}\right)^{a_3} \cdot e^{-\frac{E_{emi}}{kT}} \quad [2.17]$$

$$\frac{1}{\tau_{OFF}} = C_4 \cdot \left(\frac{I_{Sub}}{W}\right)^{m_2} + C_5 \cdot \left(\frac{I_{ds}}{W}\right) \cdot \left(\frac{I_{Sub}}{I_{ds}}\right)^{m_3} \quad [2.18]$$

where C_1 (SVE), C_2 (EES), C_3 (MVE), C_4 (off-state) and C_5 (sub-V_T) are all the subsequent constants determined in their respective dominant mode. This modeling has been validated for various T_{ox} = 5 nm, 3.2 nm and 1.7 nm, under a large set of voltage conditions (VGS, VDS), temperatures and device geometries in both NMOSFET and IO PMOSFET.

2.2.3.6. *Risk analysis method applied to DDR3–FF20 (manufacturer A)*

This part copes with the memory analysis proposed by manufacturer A for data storage as an SDRAM circuit processed on a 20 nm FinFET (FF20) technology. The first section aims to determine the sensitive parameters that have an accelerating effect on wearout and failure modes. The second section will propose the extraction of a generic structure from CA (see Figure 2.7, step 1B) at the frontend, while the third section concerns the lifetime determination using the chosen accelerated failure mechanisms (see Figure 2.7, steps 2A–2B) according to the referred modeling from the literature and the mission profile (MP). Finally, a strategy of data exploitation will be given in the fourth section with respect to AC operation.

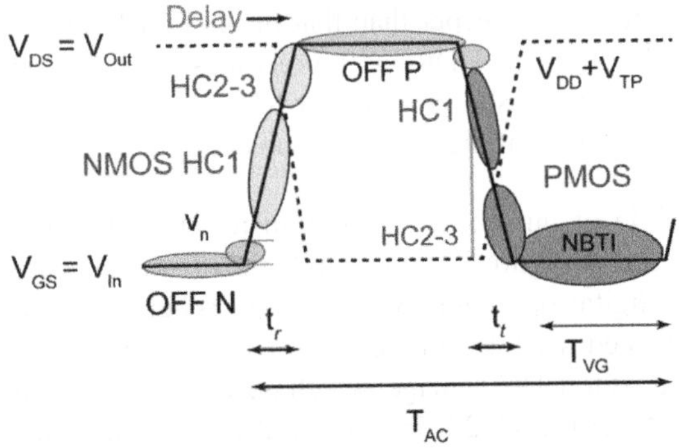

Figure 2.15. *Inverter-like pulse waveform used for the DC–AC transfer of each accelerating degradation rate in a three-mode energy modeling devoted to core and IO CMOS nodes. The off-mode phases are included both in NMOS and PMOS for the use of QS device lifetime extrapolation (Bravaix et al. 2010). For a color version of this figure, see www.iste.co.uk/elhami/mechatronics.zip*

2.2.3.6.1. Extraction of the sensitive technology parameters

This technology is developed from the recent FinFET 20 nm and 16 nm processing developments, but they have an originally distinct application scheme. Thus, it is necessary to consider the main differences regarding the general application type of the mentioned component and extract the main parameters and operation conditions given by the manufacturer that are involved in the wearout damage. This will offer the fundamental parameters required for the lifetime modeling.

2.2.3.6.2. DD3R–FF20 general temperature and supply voltage characteristics

The DDR3 is a synchronized dynamic random-access memory (DRAM) by an external clock, so the central operation type is related to DRAM semiconductor memory functioning that can run at faster speeds than conventional DRAM. So, the synchronization is done at a double rate linked to the leading (CK) and falling edges (CK#) of the clock cycle, where this memory type is volatile, which means that when once the power is removed, or a supply voltage brought by degraded metallization falls, it loses all data.

The component DDR3 is referred from manufacturer A and means that the part under analysis has a data rate referred to as 1866 for a density of 4 Gb (4x1Gb) and the memory is in fact a DDR3L SDRAM with 256Mb x 16. The package contains 96 balls 8x14 mm FBGA.

The aimed applications are for the automotive market with a temperature range of -40°C ≤ TC ≤ 120°C, and for the industrial (commercial) market with TC = -40°C to 95°C. TC is the maximum temperature measured at the center of the package. The junction temperature is given by the data sheet (automotive) as [4Gb_DDR3L]:

$T_{Junction}$ (°C)	SLOW	TYPICAL	FAST
Commercial	110	50	-40
Automotive	120	50	-40
Automotive UH	140	50	-40

The timing cycle time for SDRAM 1866 is T_{AC} = 1.07ns with programmable CAS (READ) latency CL = 13 corresponding to a frequency F_{AC} = 934 MHz.

The given supply voltages are V_{nom} = 1.35V with V_{DDmin} = 1.283V and V_{DDmax} = 1.425V, which is backward-compatible to V_{DD}= 1.5V ± 75mV application with DDR3 SDRAM. The same supply voltages are supplied for internal logic (V_{DD}) and for the data bus IO pin drivers (V_{DDQ}).

This allows us to extract the spice netlists for automotive constraints as [4Gb_DDR3L]:

LIB	Data rate	Process	Die Cap	V_{DD}	Temperature
***SLOW_1066_AAT**	1,066–1,866 Mbps	Weak	Maximum	1.283 V	120°C
***TYP_1066_AAT**	**1,066–1,866 Mbps**	**Typical**	**Typical**	**1.350V**	**50°C**
***FAST_1066_AAT**	1,066–1,866 Mbps	Strong	Minimum	1.425V	-40°C

The main block to be considered is the memory arrays with a sense amplifier and with IO gating with mask logics and a column decoder. As we want to pick up a generic structure representative of the DDR3 unitary cell, we first have to extract the main elements, helped by CA, of the double data rate synchronous RAMs (burst mode), which achieve a high-speed operation.

2.2.3.6.3. DD3R–FF20 structural characteristics

The general operation of the DRAM block to the DRAM cell is illustrated in Figure 2.16, where a row decoder selects the word line (WL) x_i of the row in which the addressed word (or cell) is located. The row decoder is also known as an x-decoder. A column decoder selects the bit line (BL) select line y_j of the column, in which the addressed word (or cell) is located. Address buffers connected to the address inputs drive the row and column decoders. A sense amplifier detects the contents of the selected cell via the complementary bit line b_j and its complement, and data bus line and its complement.

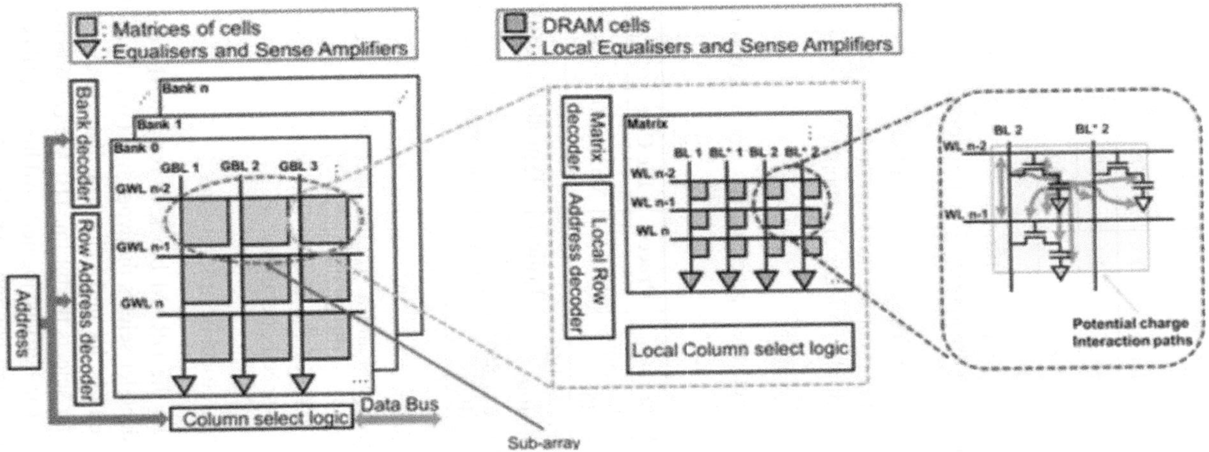

Figure 2.16. *Illustration of the DRAM block as a matrix of elementary cells and the potential charge interaction between the storage capacitors and active transistor. For a color version of this figure, see www.iste.co.uk/elhami/mechatronics.zip*

The DRAM cell is generally fabricated with trench capacitors in planar technology, but is fabricated with stacked (3D) capacitors for higher-density RAM. Here, the DDR3–FF20 component is composed of an innovative three-dimensional cell structure, such as trench, and stacked, which makes its realization possible with much higher-density DRAM cells up to 4 Gb. For either trench or stacked structure, the capacitance had to be boosted without increasing the projected area (or even decreasing the area) with increasing DRAM density.

From the CA report, it is important to identify and distinguish the various elements:

– Vertical capacitors are localized between M2 and poly layers. Capacitor top electrodes contact each other to form a common top plate. This top plate is composed of Ti.

– Below this common top plate, vertical capacitors seem to have a doubled structure. They are formed by successive concentric tubes of a SiTi top electrode (inner tube), a ZrO dielectric, a Ti intermediate electrode (intermediate tube), a ZrO_2 dielectric, and a Ti bottom electrode (outside tube). The intermediate electrode is connected to a single active area. A de-processing has been performed to confirm this doubled structure capacitor.

– Inside the memory table, metal M1 cubes are used to connect capacitor intermediate electrodes to transistor active areas. Metal M1 cubes are made of W deposited on Co (W/Co).

– Dielectric ILD0 is a SiO single layer (SiO) that is supposed to be assimilated to SiO_2.

– The poly layer is localized below the capacitors. Poly lines are covered by two silicides (SiW on SiTi) and a SiNC cap. Inside the memory table, the minimum (drawn) poly length is measured at LG $\approx$ 20 nm.

– Inside the memory table, poly cubes are used to connect the capacitor intermediate electrodes to the transistor active areas.

– Inside the memory table, an unusual metal M0 layer is present below the poly layer. Metal M0 is made of W deposited on a bottom Ti barrier.

2.2.3.6.4. Generic DRAM cell for DDR3L–FF20

From the CA analysis, the DRAM cell exhibits a 3D structure with capacitance storage on top and 3D transistors at the bottom (see Figure 2.17) controlled by M0 lines that drive the bit line (BL) on their drains, and poly lines that drive word line (WL) on their gates. The example in Figure 2.17 gives a DRAM cell, but with trench capacitor storage. Beyond 50 nm, only stacked-capacitor memories are manufactured. Hence, in our case, the capacitor seems to be a double-sided cylinder capacitor with a Ti top plate M2 electrode and a Ti bottom electrode to poly gate. This has to be examined in detail as high-K dielectric (ZrO_2) is used in order to increase the stored charge capability.

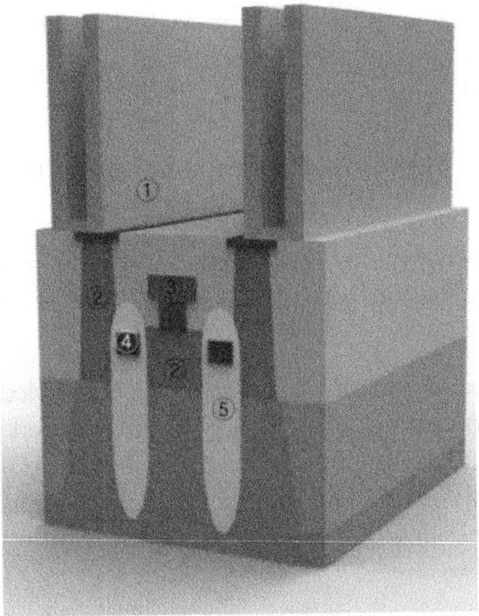

a) The schematics of a DRAM cell with the storage capacitors (① Capacitor, ② SAC, ③ Bit line, ④Word line, ⑤ Gate)

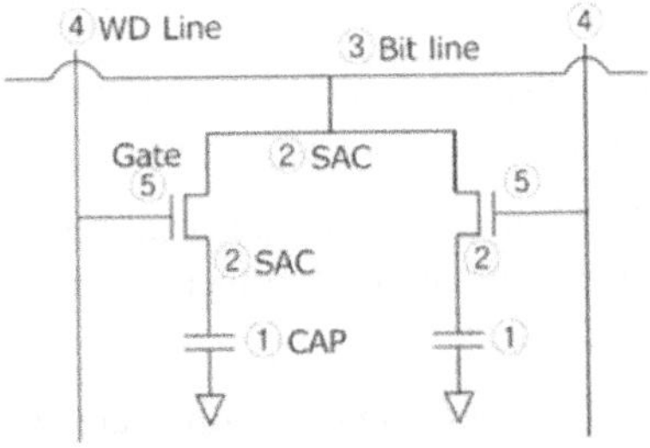

b) Equivalent circuit of the DRAM cell structure

Figure 2.17. *Analysis in the DRAM cell structure (Shin et al. 2013). For a color version of this figure, see www.iste.co.uk/elhami/mechatronics.zip*

Vertical storage capacity

In order to produce a dynamic random-access memory 20 nm DRAM or less, the most important concern regarding operation is:

– to reduce the leakage current degradation of the capacitor using high-k dielectrics eliminated by removing external impurities of boron and hydrogen without any change in the structure or materials of the capacitor: what seems to be clear is the storage capacitor (see Figure 2.17);

– to implement a double-sided capacitor in that dimension. Stacked capacitor DRAMs are built from twin cells, meaning that two cells share the same bit line contacts. Trench-capacitor areas (8 F^2/bit) are therefore usually larger in area than stacked-capacitor cells (6 F^2/bit), with F being the half-pitch (in our case, F = 21 nm). Because of the use of vertical pillar transistors (VPT) (drain stacked on top of the source (see Figure 2.18), separated by the channel length), a single memory cell area is expected to reduce to 4 F^2/bit.

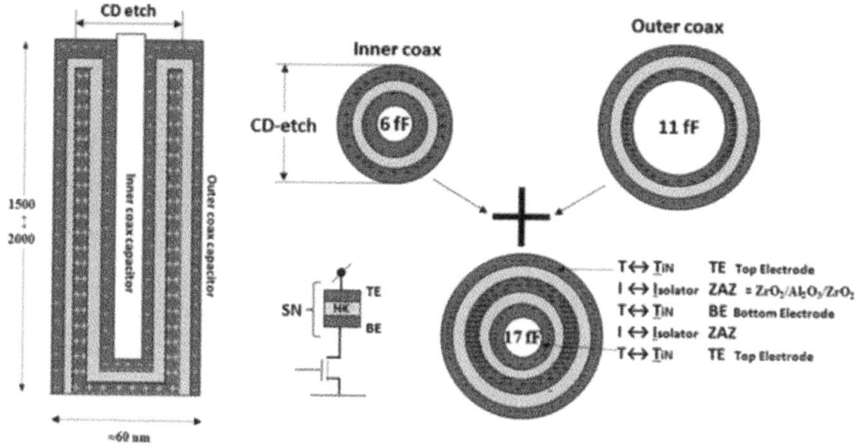

Figure 2.18. *Double-sided charge capacitor referred to as a crown capacitor for DRAM cells [(Veendrick 2017) source Vreugdenhil; ASML semiconductors]*

In Figure 2.18, we show an example that was used (in 2013) in 25 nm DRAM technology (Veendrick 2017) This capacitor represents a capacitance value of 17 *fF*, which is built from two cylinders. The inner and outer cylinders represent, respectively, 6 *fF* and 11 *fF* of storage capacitance. This tri-layer high-K dielectric ZAZ (see Figure 2.4) has been processed as a stack of zirconium–aluminum–zirconium oxides giving a thickness td of approximately 9 nm. We have the same kind of structure cell in DDR3, except that only one material is used for the high-K.

From CA extraction, there is only a single dielectric material ZrO(2) identified as the high-K dielectric:

– we have a height $H = 882$ nm and Si mesh of height $h = 615$ nm;

– the inner coax tube is in SiTi;

– there is high-K in ZrO_2, with a dielectric constant $\varepsilon_R = 40$ (to 45) for $T_{ox} =$ 6 (to 9 nm) respectively.

This simplifies the structure in DDR3L as:

– the inner coax Si/Ti with a thickness of 18.42 nm (first diameter = $2R_1$);

– the successive tubes of ZrO/Ti/ZrO with a thicknesses of 6.14 nm/ 6.14 nm/6.70 nm;

– the inner electrode and the outside electrode are connected to the common top plate.

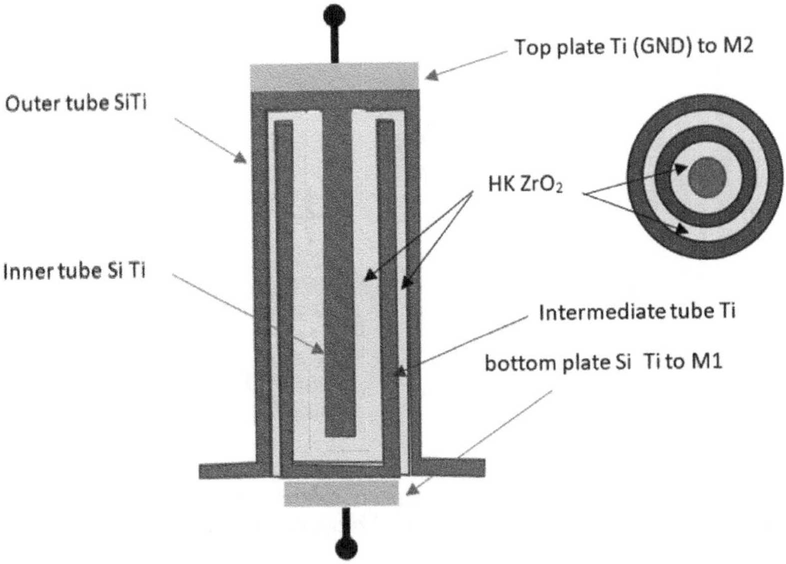

Figure 2.19. *Schematics of the double-sided capacitor for DDR3–FF20. For a color version of this figure, see www.iste.co.uk/elhami/mechatronics.zip*

This tubular cell capacitance (see Figure 2.19) is known to be much more robust in terms of process variability. One disadvantage of the cell is its larger capacitor area, but its aspect ratio is 3.5 times smaller, allowing for faster DRAM operations. This crown capacitor cell has been used since 2015 for the 20 nm node DRAM cell in DDR3 and DDR3L.

For our HK thickness range, the dielectric constant of zirconium oxide (ZrO2) is approximately 45–35 depending on its amorphous phase, or between 30 and 20 as a crystalline phase when it is mixed as in ZAZ. Below a thickness of 5 nm, its value dramatically reduces with thickness. CA gives a 6.1–6.7 nm thickness. We will choose a medium value $\varepsilon_R = 40$.

The cylinder capacitance is $C = 2\pi.\ \varepsilon_o.\varepsilon_R\ .H/Ln(R_2/R_1)$ taking into account $\varepsilon_R = 40$, the height H and R_1 and $R_2 = 15.3$ nm measured from the CA. This gives $C_1 = 1.96\ fF$ for the inner capacitance. For the outer capacitance, we have $C_2 = 7.23\ fF$. Then, for the double charge (parallel) capacitance, we have: $C_{tot} = 9.19\ fF$.

The bottom of the storage capacitor is connected to the cell transistors thanks to M1 cubes. We observe ILD1 in SiO2, M1 level with WCo, then silicide and poly/SiO, then M0/gate oxide and active + STI in the shape as "Saddle FinFET".

This means that the 3D FF has been rotated by 180°, giving an inverse gate control with respect to the fin oriented between S/D.

3D cell analysis of DDR3L

In the DDR3L mode of operation, the DDR3 bus rate is given at 1.5 GHz (800–1,600 MHz) at a maximum of 1.5 V. The bandwidth of DDR3 is doubled compared to that of DDR2, with no major speed change of the DRAM core, that is, the memory cell array. To achieve this double bandwidth, DDR3 uses a prefetch of eight words instead of a four-word prefetch for DDR2. This means that for every single read or write operation, a total of eight words are accessed in parallel (shown by the parallel fins) in the DRAM core to enable the high data rate at the interface.

Thanks to the CA study (Hirex Engineeering 2019), the 3D core cell has been identified, showing that this technology uses FinFETs with buried word line. The active in the cell matrix area is disorientated at 30° shift versus bit line and word line. Each active block is isolated from the adjacent one with STI oxide.

On each active island, two buried word lines, two capacitor contacts and a single central bit line contact are observed. Pictures are shown in the schematic of Figure 2.20.

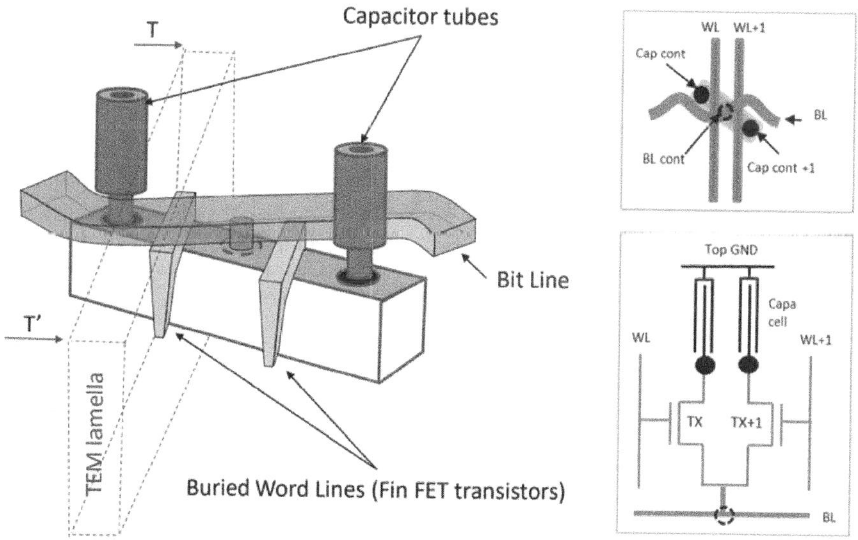

Figure 2.20. *Schematics of the cell structure synthesis obtained from the TEM lamella axis showing the double-sided capacitors and paired transistors connected to WL and WL+1 (Hirex Engineeering 2019). For a color version of this figure, see www.iste.co.uk/elhami/mechatronics.zip*

With a decreasing DRAM cell size, a recessed channel array transistor (RCAT) has been proposed to overcome the short channel effect (SCE) of conventional MOSFETs with planar channels. To use the fabrication process simplicity of planar cell array transistor, a vertical transistor, RCAT structure was given in stack capacitor cells. The basic concept of the RCAT is to increase the effective length (L_{eff}) by recessing the channel from the silicon surface. So, channel doping density can be reduced, thereby reducing the S/D resistance of the memory cell transistor, and enhancing carrier mobility, in addition to the main advantage of reduced leakage.

Although the recessed channel of RCAT has improved the short channel effect (SCE), RCAT suffers from low driving current and V_{TH} sensitivity due to the shape of the bottom corner of the recessed channel (Shin et al. 2013). To solve these problems, a saddle FinFET (S-FinFET) has been proposed with a tri-gate that wraps both the recessed channel surface and the side surface (Crupi et al. 2013; Kim et al. 2019), as shown in Figure 2.21. S-FinFET not only exhibits excellent short channel effect characteristics, but also maintains excellent subthreshold swing (SS), high ion, and nearly constant V_{TH}.

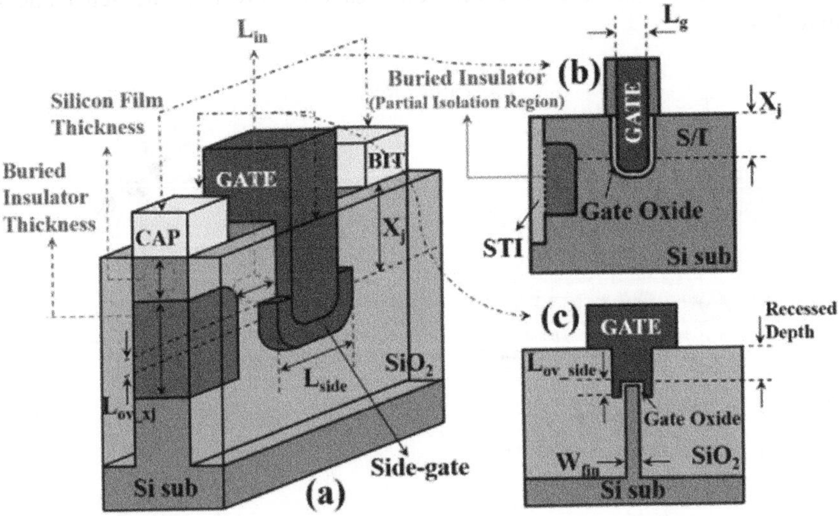

a) 3D schematic of Pi-FinFET b) Cross-sectional view across the gate; c) Cross-sectional view of the thin body

Figure 2.21. *3D schematic of Pi-FinFET (Crupi et al. 2013; Kim et al. 2019)*

The gate wraps three surfaces of the recessed channel, similarly to a FinFET. The buried insulator material is used with SiO_2 below the storage node. The buried insulator is penetrated from the shallow trench isolation (STI) region. X_j of the source/drain (S/D) is located 112 nm from the top surface of the S/D region with a Gaussian profile. The peak concentration of the Gaussian S/D doping profile is $1.5 \cdot 10^{20}$ cm^{-3}, and the uniform body doping concentration is $5 \cdot 10^{17}$ cm^{-3} (Kim et al. 2019).

2.2.3.6.5. Extraction of the generic structure for DD3RL

The complete analysis of the SDRAM array shows that the full cell matrix is very complex in design and processing at that level of integration and memory density. Many uncertainties still remain, although the CA study has led to a number of improvements and helps in determining the nature of materials and dimensions. This leads us to simplify the approach to a generic (simpler) digital cell based on the CA dimensioning. As the objective is to consider a generic structural device to which a reliability methodology can be applied, leading to the determination of an "intrinsic reliability" examination with failure rate λ_G, we will consider that the cell is composed of paired N-channel transistors very similar to FinFETs wrapped around the recessed channel, with buried WL connected to the gates of the FinFETs and BL connected to the (common) drain to reach the required drivability of the cell.

Figure 2.22 shows a schematic of a matrix 3D cell of the DRAM with paired transistors and double-sided cylindrical (coax) capacitors (22 nm node). In the periphery CMOS region (interface area), we have an inter-block area between M1 in the inter-block and M0 in memory cells, with standard planar transistors (isolated by STI) with poly gates and thicker gate oxide (T_{ox} = 2.8 nm).

Consequently, we propose that for DDR3L, the generic structures can be chosen as a memory point composed of N-channel FinFET with its advanced processing (RCAT, recessed channel, saddle FinFET) and storage capacitance, and the other case of standard planar transistors. This will provide two points of examinations:

– One with a very short channel representing the worst-case reliability with respect to FEOL at the length of the technology node (T_{op} part L_{eff} = 16 nm).

– Another one with relaxed design rules for periphery CMOS with planar process (poly gate and STI isolation) at a longer device length (L_{eff} = 102 nm).

The CA study parameters can be summarized in the table below, with a summary of the physical process parameters obtained for memory cell and peri standard CMOS:

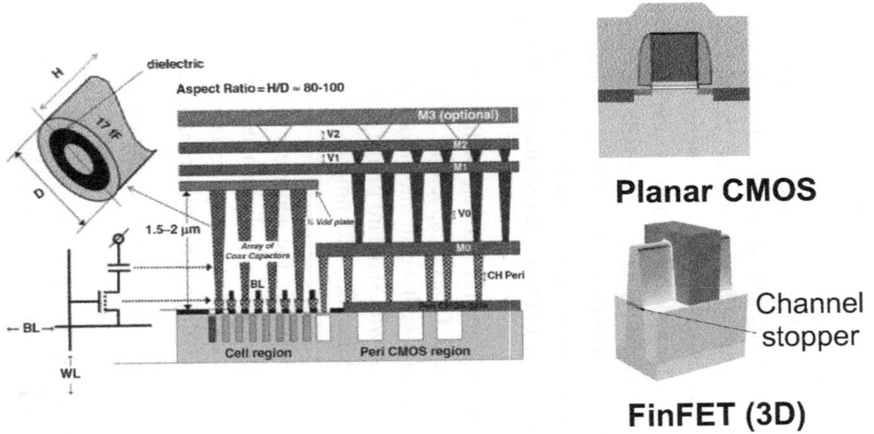

a) Schematics of the matrix 3D cell of the 22 nm DRAM with paired transistors and double-sided cylindrical (coax) capacitors (22 nm node)

b) Extractions of 3D FinFET (DRAM) and planar (Peri) CMOS

Figure 2.22. *Architecture for SDRAM (Veendrick 2017). For a color version of this figure, see www.iste.co.uk/elhami/mechatronics.zip*

Device type	WxL drawn	L_{nom}	L_{eff}	T_{ox} = EOT	V_{DD} (V) min/Typ./Max
FinFET in SDRAM (3D)	33 x 37nm	**20 nm**	**16 nm**	2.8 nm	1.283/**1.35**/1.425 (1.5 V)
Std peri CMOS (planar)	33 x 37nm	**20 nm**	**16 nm**	2.8 nm	1.283/**1.35**/1.425 (1.5 V)

Note that for the FinFET, we can check the rule in width, where $W_{eff} = 2\ H_{Fin} + W_{Fin}$ to determine the effective channel width with respect to the drawn width, and if we do observe multi Fins in parallel to the pictures, we have to account for an effective width of $W_{eff} = [2\ H_{Fin} + W_{Fin}] \cdot N_{Fin}$ that would give for a supposed $H_{Fin} = 50\ nm$:

W_{eff}= [2·50 nm + 33 nm] = **133 nm** per fin.

2.2.4. *Reliability study for DD3RL*

In this section, we study the SDRAM reliability under the worst case of aging, which we analyze through the reliability under TDDB, HC and NBTI in a first step and the CA reports about that device. We have to use the FinFET dimensions for DRAM cell at nominal L_{nom} and a single effective W_{eff}, that is, for one fin width, extracted from the DPA for the core N-channel FinFET of the memory cell.

2.2.4.1. *TDDB reliability*

The TDDB reliability of the core cell can be determined by the 22 nm results from the literature on similar FinFET from other foundries (INTEL), where the dimensions are very close to DDR3 dimensions (Ramey et al. 2013) (fin height $H{\sim}34\ nm$, fin width $Wsi{\sim}8\ nm$, the gate length is drawn at 30 nm and gate pitch 90 nm, HKMG stack). Ramey et al. (2013) indicate that the threshold voltage of FF is $V_{TH} = V_{TH}$ (planar) – 0.1V in that case. The TDDB modeling is done by the E-model which is the most restrictive. The exponential dependence gives the following parameter usually required for TDDB, but the lack of a reference area (A_G) and TDDB distribution in the published results (Ramey et al. 2013) avoids the AF_{AG} factor evaluation. Based on the beta value, we are able to define the area scaling by this relationship:

$$TTF_{BD1}/TTF_{BD2} = \left(A_{G2}/A_{G1}\right)^{1/\beta} \qquad [2.19]$$

where the Weibull parameter β ranges from 1.1 to 1.7.

So with this assumption, we found that using the direct extrapolation from the results (see Figure 2.10) for DDR3 FF with $L_G = 20\ nm$ at typical 1.35 V and maximum 1.5 V operation gives: $TTF_{BD} \propto A_i \,.\, exp(-\gamma V_G) \,.\, exp(-E_a/kT)$.

$\tau_{TDDB} \propto$ exp (-γ E) model	NMOS	PMOS
$V_{DD} = 1.35$ V	$5.22\ 10^9$ s = 165 years	$6.65\ 10^{11}$s = $2.1\ 10^4$ years
$V_{DD} = 1.5$ V	$2.58\ 10^7$ s = 0.81 years	$4.21\ 10^{10}$ = $1.33\ 10^3$ years

This shows (with the aforementioned assumption) that the DDR3 at a typical value of 1.35 V is well above the lifetime criterion for PMOS with no TDDB sensitivity, but the NMOS exhibits a larger acceleration factor with V_{DD} that puts the boost condition at $V_{DDmax} = 1.5$ V, the most sensitive factor with respect to TDDB.

2.2.4.2. *HC reliability*

As stated before, the difficulty in extrapolation lifetime data is that we do not know the process dependent constant value C_i yet, which differs from batch to batch and from silicon bulk to FinFET CMOS node. What can be done first is to determine the lifetime value from I_{On} performance from the single transistor or from a block of (idealized) identical geometry, because the I_{On} current per device is much smaller in this FF 20 nm process, which has to be analyzed for its typical nominal supply voltage $V_{DD} = 1.35$ V and (a) for EOT = 2.8 nm and (b) $L_G = 130$ nm ($L_{eff} = 102$ nm) for EOT = 1.5 nm.

2.2.4.2.1. Short channel LG = 20 nm from the memory point

A first direct evaluation could be done by considering the average current delivered by the memory block DDR3 – 1866 giving 12 mA/16 banks = 750 μA per bank as row addressing denoted by A[n:0] for 4 Gb is: n = 14 (x16).

This leads to a level of nominal current at I_{On}= 750 μA/μm for a lifetime calculation (see Figure 2.13) as $\tau_{HC} = C_i\ (V_{DS}^{0.5} \cdot I_{DS})^{-24.3}$, which provides the relationship for C_i calculation. This may be related directly to the experimental lifetime plot of the (close) 14 nm FF node, giving a DC lifetime (Chang et al. 2019) as a function of:

$$I_{On}\ (\mu A/\mu m) = I_{DSat}\ (V_{DD}, V_{DD})\ /\ W_{eff} \qquad [2.20]$$

This makes the nominal I_{On} the reference value, giving $I_{DS} = 99.75\ \mu A \cong 100\ \mu A$. Due to the quadratic dependence on $(V_{GS} - V_{TH})^2$, the value of $I_{DS}(V_{DD2})$ can be obtained assuming identical V_{TH} increasing V_{DS} (with negligible drain-induced barrier lowering effect if V_{DD} is increased to 150 mV) as:

$$I_{DSat1}(V_{DD1})\ /\ I_{DSat2}\ (V_{DD2}) = (V_{DD1}\text{-}V_{TH1})^2 /\ (V_{DD2}\text{-}V_{TH2})^2 \approx (V_{DD1}/V_{DD2})^2,$$

giving: $I_{DSat2} \approx I_{DSat1}\ (V_{DD1})\ (V_{DD2}/V_{DD1})^2 = 1.291\ I_{DSat1}\ (V_{DD1})$

thus, leading to a 30% increase,

$W_{eff} = 0.133\ \mu m$ **per fin** $I_{On} = 750\ \mu A/\mu m,\ V_{DD1} = 1.35\ V$	$I_{DSat}\ (V_{DD}) \cdot V_{DD}^{0.5}$	**Lifetime (s)**
$ID_{Sat1} = 100\ \mu A,\ V_{DD1} = 1.35\ V$	$1.16 \cdot 10^{-4}$	$1.33 \cdot 10^{17}$
$ID_{Sat2} = 129\ \mu A\ V_{DD2} = 1.50\ V$	$1.58 \cdot 10^{-4}$	$3.24 \cdot 10^{14}$

This very large lifetime value for DC indicates that the HC does not seem to be an issue at such low voltage and small dimensions.

If we now consider two and four fins width in parallel, for example, in an inverter, results will be quickly changed, but the power law I_{On} *(1 Fin)* ⇨ $I_{DSat} = I_{On} \cdot W_{eff} \cdot N_{Fin}$ translates the results with respect to the x-coordinate:

$W_{eff} = 0.133\ \mu m$, *Inverter x2Fins* $I_{On} = 750\ \mu A/\mu m,\ V_{DD1} = 1.35\ V$	$I_{DSat}\ (V_{DD}) \cdot V_{DD}^{0.5}$	**Lifetime (s)**
$ID_{Sat1} = 200\ \mu A,\ V_{DD1} = 1.35\ V$	$2.320 \cdot 10^{-4}$	$1.46 \cdot 10^{11}$
$ID_{Sat2} = 258\ \mu A\ \ V_{DD2} = 1.50\ V$	$3.159 \cdot 10^{-4}$	$2.662 \cdot 10^{8}$

$W_{eff} = 0.133\ \mu m$, *Inverter x4Fins* $I_{On} = 750\ \mu A/\mu m,\ V_{DD1} = 1.35\ V$	$I_{DSat}\ (V_{DD}) \cdot V_{DD}^{0.5}$	**Lifetime (s)**
$ID_{Sat1} = 400\ \mu A,\ V_{DD1} = 1.35\ V$	$4.647 \cdot 10^{-4}$	$1.03 \cdot 10^{5}$
$ID_{Sat2} = 516\ \mu A\ \ V_{DD2} = 1.50\ V$	$6.319 \cdot 10^{-4}$	$2.46 \cdot 10^{2}$

Then, it is clear that the FF 20 nm technology is limited in the number of fins for the current drivability optimization, as the lifetime strongly decreases as a function of two fins per device for $V_{DD}=1.5\ V$ (< 10 years) and for supply voltage if we put four fins in parallel.

As these last cases would correspond to the case of a digital inverter, it can be translated to the real AC operation condition, taking into account the time for pull-up (and pull-down) of the equivalent stage as a function of fan out (FO).

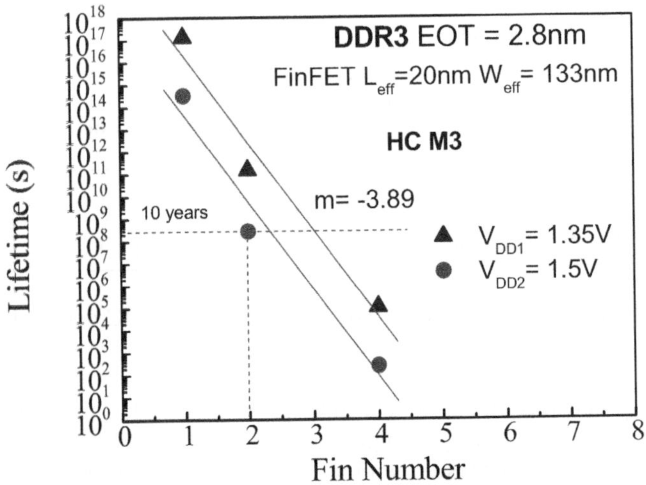

Figure 2.23. *Summary of the HC (DC) lifetime in DDR3 as a function of the number of fins for NMOS inverter side for equivalent N-channel FinFET transistor. For a color version of this figure, see www.iste.co.uk/elhami/mechatronics.zip*

The M3 mode effect is enhanced during the transient through the age increase when tr is reduced from 500 ps to 100 ps. A similar AC behavior has been demonstrated in PMOSFETs (Bravaix et al. 2009), as both CHC N- and P- MOSFETs worst-case degradation have moved to the high V_{GD} region. This transfer to our AC condition allows us to define HC design guidelines due to the assignments of main degradation in M2 to M3 modes, which can be merged to M3 only for low voltage operation. The reduction of *NAF* (*PAF*) factors, which are defined here as a function of AC–DC time factors $NAF= NTF{\cdot}f_{AC}$ with $NTF = x_n/(f_{AC}{\cdot}t_r)$, where x_n is the proportion of the CHC to CCC damage in the rising transient, that is, with $T_{hot,n} = t_r / x_n$ and $T_{hot,p} = t_f / x_p$ with the corresponding relationships looking at Figure 2.24.b:

$$TF = \frac{T_{AC}}{T_{Hot}} = \frac{1}{f_{AC}\left(t_{r,f}/x_{N,P}\right)} \Leftrightarrow NTF = \frac{x_N}{f_{AC}{\cdot}t_r} \text{ and } PTF = \frac{x_P}{f_{AC}{\cdot}t_r} \quad [2.21a]$$

$$T_{Hot}(NMOS) = \frac{t_r}{x_N} \text{ and } T_{Hot}(PMOS) = \frac{t_r}{x_P} \quad [2.21b]$$

which are defined at shorter $t_{r,f}$ and larger fan out (FO), in relation to the larger impact of M3 mode, showing a trade-off to optimize between frequency f_{AC}, t_r, t_f and the FO.

We can simply the input signal time between V_{TH} and $0.9 \cdot V_{DD}$ over x factor that can be obtained, as for FO = 1 we have $t_{r,f} = 0.03 \cdot T_{AC} = 32\ ps$, so that we can be closer to real AC stressing configurations at the device level. Then, the effective (M3) damaging mode could be simplified to $0.9 \cdot t_{r,f}/2$ duration, which enables us to calculate the NTF (DC-AC) factor, giving:

$$\text{NTF} = 1.07\text{ns} / (0.9 \cdot 32\text{ps} / 2) = 74.30 \cong 75 \text{ for typical estimate}$$

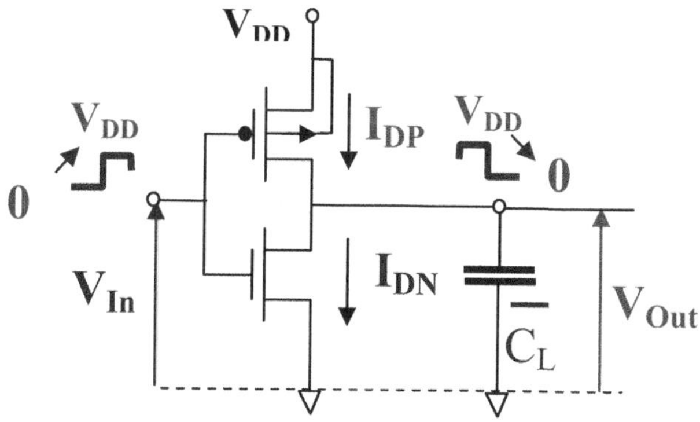

a) CMOS inverter operation schematics as a function of V_{In}, V_{Out} and its load capacitance C_L

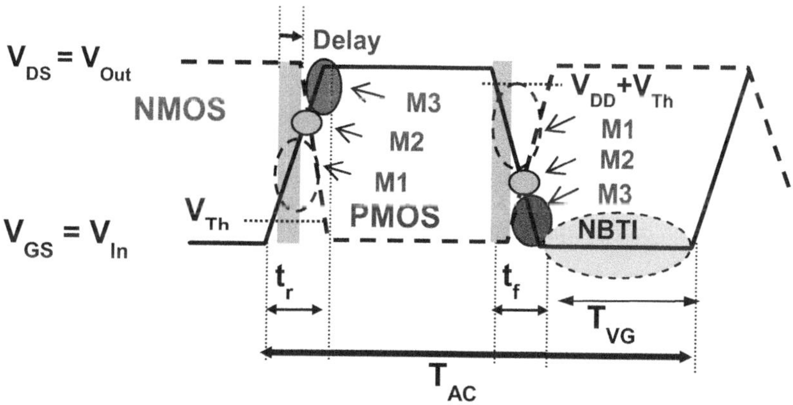

b) Signals applied to the inverter $V_{Out} = f(V_{In})$ related to the distinct bias region where HC to CC are involved.

Figure 2.24. *CMOS inverter operation schematics (i) for the NMOS transistor, where $V_{IN} = V_{GS,N}$ and $V_{OUT} = V_{DS,N}$, and (ii) for the PMOS transistor, where we have $V_{GS,P} = V_{IN} - V_{DD}$ and $V_{DS,P} = V_{OUT} - V_{DD}$ due to the inverted potential reference for the PMOS source with $V_{S,P} = V_B = V_{DD}$ (Bravaix et al. 2009). For a color version of this figure, see www.iste.co.uk/elhami/mechatronics.zip*

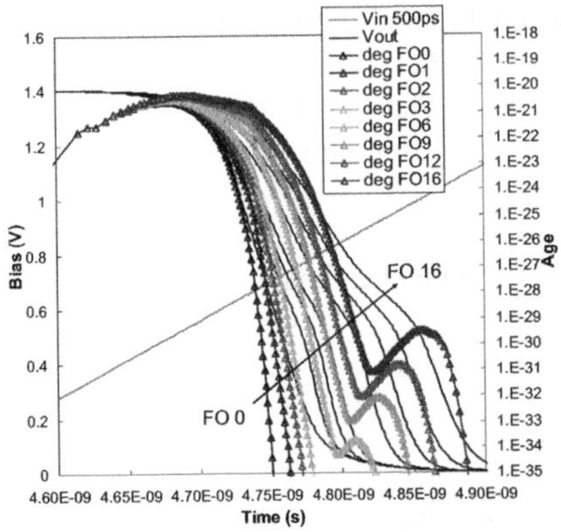

a) For tr = 500 ps

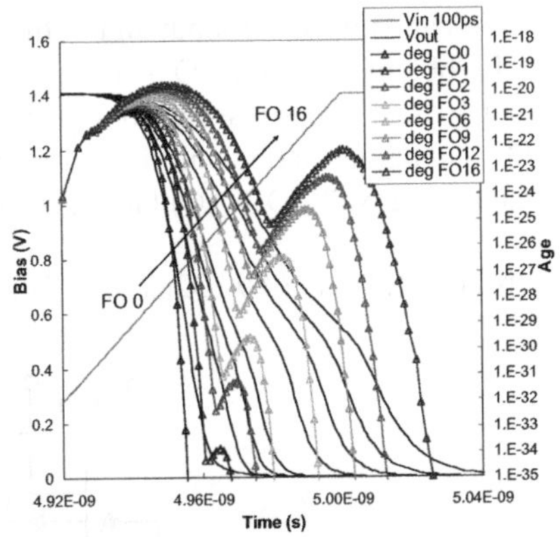

b) for tr = 100 ps

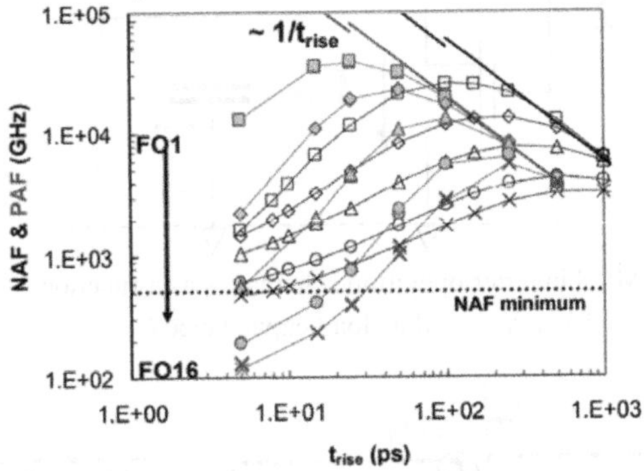

c) NAF, PAF (full, red) versus tr with FO variable in core inverters showing an NAF (and PAF) fall down with tr

Figure 2.25. *Function age defined as age(t) = t_s/τ in a 40 nm NMOS inverter as a function of FO = 1 to 16 (Bravaix et al. 2009). For a color version of this figure, see www.iste.co.uk/elhami/mechatronics.zip*

This consequently relaxes the AC lifetime for two fins case, at V_{DD} = 1.5 V which becomes higher than the lifetime criterion (10 years) as we may apply:

$$\tau_{AC} = \text{NTF} \cdot \tau_{DC}$$

But the four fins case remains below the 10 year lifetime criterion independently of V_{DD}.

2.2.4.2.2. Long-channel LG = 130 nm from the standard peri CMOS

The same approach is applied to DC lifetime evaluation in L_G = 130 nm (L_{eff} = 102 nm, with thinner gate oxide EOT = 1.5 nm), which can be considered as "long-channel" behavior with respect to the one with L_{eff} = 16 nm if we know the current value.

In this case, the lifetime slope exponent reduces from -20 to -17.9 (Huard 2010), but would deviate from the M3 mode due to long-channel dependence and smaller HC damage, even for this thin EOT. Then, as we do not know the I_{On} (μA/μm) or channel width for the peri CMOS, we can apply a standard lifetime technique coming from the classical HC, in order to include the accelerating L_G dependence which is sensitive for that long-channel case, as well as for IO devices that operate at higher voltages.

Indeed, assuming a time power law for the reduction of the I_{On} current (Δ_{IDSat} = *10% FWD* = Δ), the normalized failure criterion follows $\Delta_{IDSat} = A\ t_s^{\ n}$, where exponent n = 0.4, we can express the lifetime:

$$\tau_{HC,\,Long} = C_i\,(\Delta / 0.1)^{1/n} \cdot L_G^{\ m} \cdot \exp(B(1/V_{DS} - 1/V_{DD}))$$

Calculated for L_G= 130 nm, m = 10.7, $C_i = 1.81 \cdot 10^5$ years = $5.708 \cdot 10^{12}$ s, B = 80 (long channel), V_{DD}= 1.35 V (typical) to maximum V_{DS}= 1.5 V, giving:

$\boldsymbol{\tau_{HC}}$ = **5.03 years** in DC

As a result, the pull-up of a block (device) designed with the FF 20 nm DDR3 would result in an AC lifetime at:

$\boldsymbol{\tau_{AC}}$ = NTF · τ_{DC} = 75 · 5.03 years = **377 years (AC)**

This shows that with many assumptions, the standard lifetime (AC) value may, in any case, satisfy the 10-year lifetime for the inter-block standard NMOS transistor of the SDRAM cell at V_{DD} = 1.35–1.5 V (typical of the maximum value). With the same assumed frequency for operation and shape of the pulse for FO = 1, this leads to an AC lifetime for L_G = 130 nm at 1.35 V, τ_{HC} = 377 years in AC (ΔI_{DSat} = 10%).

2.2.4.3. *NBT reliability*

Very few results show quantitative TTF under the NBT mode for FinFETs. Most of the results are on SOI FinFETs. As we have body-tied FinFETs for the DDR3 component, we can use the acceleration modeling presented in the previous section (see Figure 2.14) with the normalization trick, with respect to the lifetime criterion (ΔV_{TH} = 50 mV). Using **[2.15]:**

$$TTF_{NBT} = A_i \cdot \left(\frac{\Delta}{0.05}\right)^{1/n} \cdot e^{-B \cdot V_G} \cdot e^{-\frac{E_a}{kT}} \qquad [2.22]$$

with:

Core	IO
$n = 0.16$ (core $\Delta V_{TH} = 50$ mV)	$n = 0.15$ ($\Delta ID_{Sat} = 10\%$)
$A_i = 1.48\ 10^8$, $B = 12$	$A_i = 5.02\ 10^7$, $B = 4.6$

With apparent activations:

$E_a = 0.25$–0.35 eV (Si bulk);

$E_a = 0.55$–0.65 eV (FDSOI);

$E_a = 0.15$–0.20 eV (FinFETs).

From CA, the supposed width of a single FF is 10 nm leading to $W_{eff} = [2.H_{Fin} + W_{Fin}]$ effective width:

W_{eff}(per fin) = [2 x 45 + 10] = 100 nm.

The results presented in Lee et al. (2005) do not exhibit the NBTI issue (T = 125°C), as the extrapolated lifetime for FinFET (body-tied) per fin width shows:

For $W_{eff} = 5$ nm $V_{DD} = 1.35$V ⇨ $\tau_{NBT} = 2 \cdot 10^{10}$s = 634 years

For $W_{eff} = 5$ nm $V_{DD} = 1.5$V ⇨ $\tau_{NBT} = 4 \cdot 10^{9}$s = 126 years

For $W_{eff} = 10$ nm ⇨ $\tau_{NBT} > 2 \cdot 10^{10}$s independently of V_{DD} (1.5 V or 1.35 V)

This arises from the much smaller vertical field condition for the DDR3 FF 20 nm, as we have:

$E_{ox} \cong V_{DDmax} / EOT = 1.5/2.8 \cdot 10^{-7} = 5.35$ mV/cm

down to the minimum value giving:

$E_{ox} \cong V_{DDmin} / EOT = 1.35/2.8 \cdot 10^{-7} = 4.8$ mV/cm

In conclusion, from the investigations, we have shown that the DDR3 does not exhibit a large damage acceleration factor for the typical V_{DD} = 1.35 V operation if we base the calculation on a single FinFET with one fin. For all three damage mechanisms, no issues were found regarding the HC, TDDB or NBTI when the right design margin and accessible technological parameters were used at the minimum length and effective width (16 nm, 133 nm).

The transfer from typical V_{DD} = 1.35 V to V_{DDmax} = 1.5 V shows increased sensitivity to TDDB in the geometry of NMOS FinFETs, where the lifetime is reduced to 0.81 of a year.

Moreover, the design of a digital cell emphasizes that the number of fins has to be limited (with the current analysis of CA results) to a number of 2, as the increase at V_{DDmax} = 1.5 V (corresponding to the boost condition) makes HC sensitivity appear for two fins and more largely for four fins, where the HC criterion is no longer fulfilled for any (manufacturer) supply voltage conditions down to V_{DD} = 1.35 V (typical value).

In conclusion, we have shortly described our work devoted to a quick analysis in the context of the PFM methodology. We have projected a simplified framework analysis.

The much larger effects have been observed in the HC voltage at low voltage as it is a current driven phenomenon. Hence, the effective width per fin and the number of fins exhibit a threshold (x4), as it has a strong impact on the HC device with a cell lifetime value between 1.35 V and a boost condition of 1.5 V in the DDR3 component (Bravaix 2019a, 2019c). The HC issue is shifted from high performance to typical operating conditions when there is a larger number of fins (x6) in FPGA and a reduced supply voltage between 0.9 V and 0.85 V.

With respect to the classical TDDB and NBTI behaviors, the only analyzed results from the literature do not evidence a large lifetime reduction through these failure mechanisms as we are at much lower oxide fields. Accent has also been pointed out by the width dependence and the larger difference between the accelerating range and the field for usage. This is explained by the low voltage in the use condition, particularly for FPGA in the range of 0.9–0.85V and the enlarged oxide field condition from the accelerated (stress) voltage range for SDRAM for 1.35 V up to 1.5 V.

2.3. Aerospace electronics reliability: practical application of MTOL

We present a method for predicting the failure rate, and thus the reliability of an electronic system by summing the failure rate of each known failure mechanism. We use a competing acceleration factor methodology by combining the physics of failure for each mechanism with their effects as observed by high/low temperature and high/low voltage stresses. Our method assumes that the lifetime of each of its failure mechanisms follows a constant rate distribution and each mechanism is independently accelerated by the stress factors, which also include frequency, current and other factors that can be entered into a reliability model. The overall failure rate also follows an exponential distribution and is described in the standard FIT (Failure unIT or Failure in Time). The method combines mathematical models for known failure mechanism and solves them simultaneously at a multiplicity of accelerated life tests to find a consistent set of weighting factors for each mechanism. The result of solving the system of equations is a more accurate and unique combination for each system model by the proportional summation of each of the contributing failure mechanisms.

To this day, the users of our most sophisticated electronic systems that include opto-electronic, photonic, MEMS devices, etc., are expected to rely on a simple reliability value (FIT) published by the supplier. The FIT is determined today in the product qualification process by use of high-temperature overstress life (HTOL) or other tests, depending on the product. The manufacturer reports a zero-failure result from the given conditions of the single-point test and uses a single-mechanism model to fit an expected MTTF to the operator's use conditions. The zero-failure qualification is well known as a very expensive exercise that provides nearly no useful information. As a result, designers often rely on HALT testing and handbooks such as the FIDES, Telecordia or Mil-Handbook-217 to estimate the failure rate of their products, knowing full well that these approaches act as guidelines rather than as a reliable prediction tool. Furthermore, with zero failure required for the "pass" criterion, as well as the poor correlation of expensive HTOL data to test and field failures, there is no communication for the designers to use this knowledge in order to build reliability or to trade it off with performance. Prediction is not really the goal of these tests; however, the current practice is to assign an expected failure rate, FIT, based only on this test, even if the presumed acceleration factor is not correct.

This paradigm seems unfortunate since the manufacturers of our electronic equipment actually put in a great deal of effort and spend so much money and excellent personnel resources to learn and study each failure mechanism. Today's approach to reliability takes the intimate knowledge of the failure mechanisms, but does not communicate this knowledge downstream to the users, usually out of fear that perhaps the models or probabilistic interpretation will not be realized. Hence, the known and already characterized mechanisms that could lead to failure are left out of the communication. This leaves the final, sterilized, qualification test as the only available assessment that the user can rely on. Everyone recognizes that the resulting calculation is far from being a reliable value to predict anything about the life. Worse than that, it makes a joke of the reliability prediction process and has led to confusion at all levels. It also makes a joke of the excellent hard work of the reliability engineers who evaluate the failure probabilities and the underlying physics.

New emerging methods of reliability prediction, including probabilistic design for reliability, physics of failure, no-MTBF and other tools are being developed. These methods take many variables into consideration and look at the failure distribution, as well as multiple mechanisms and their interactions. There are new campaigns, and calls to throw away the handbooks, all of them, without at least appreciating the theoretical underpinnings that may still be valid. Then, we are left with the other debate of making deterministic assessments versus probabilistic models for reliability. However, without proper communication between the manufacturer and the user, we again ask the question of whether making predictions for reliability is possible when using models that are disconnected from the actual products.

Here, we address the pros and cons of various predictive approaches using the common language of failure in time or failure unIT (FIT). We will evaluate the goal of finding MTBF and evaluate the wisdom of various approaches to reliability prediction. Our goal is reliability prediction based on the system environment including space, military and commercial. It is our intent to show that the era of confidence in reliability prediction has arrived and that we can, in fact, make reasonable reliability predictions. There is a major value in having consistent reliability metrics and offering communication between levels of manufacturing and practical ways that upstream models can be used by downstream users to make cost-effective reliability predictions.

We have found a practical means of separating electronic device failure mechanisms at the system level. We tested FPGAs with a large range of frequency operation and at extremely high and low temperatures with voltages ranging from nominal to more than two times nominal voltages. The result is the ability to completely distinguish the influences of hot carrier injection (HCI), electromigration (EM), bias temperature instability (BTI) and time-dependent dielectric breakdown (TDDB). Our result shows that a meaningful reliability prediction can be made by a summation of distinct intrinsic failure mechanisms as measured at the system level. The result of our work will be a system qualification protocol that can actually predict the failure in time (FIT) with a much greater accuracy than a standard high-temperature or low-temperature overstress life qualification.

Chip and packaged system reliability is still measured by a failure unIT (FIT). The FIT is a rate, defined as the number of expected device failures per billion-part.hours. An FIT is assigned for each component multiplied by the number of devices in a system for an approximation of the expected system reliability. The semiconductor industry provides an expected FIT for every product that is sold based on operations within the specified conditions of voltage, frequency, heat dissipation, etc. Hence, a system reliability model is a prediction of the expected mean time between failures (MTBF) for an entire system as the sum of the FIT rates for every component.

Failures in terms of an acceleration factor, AF, as:

$$FIT = \frac{\#failures}{\#tested \cdot hours \cdot A_F} \cdot 10^9 \quad [2.23]$$

where #failures and #tested are the number of actual failures that occurred as a fraction of the total number of units subjected to an accelerated test. The acceleration factor, AF, must be supplied by the manufacturer since only they know the failure mechanisms that are being accelerated in the high-temperature operating life (HTOL), and it is generally based on a company proprietary variant of the 217^+ approach for accelerated life testing. Therefore, the true task of reliability modeling is to choose an appropriate value for AF based on the physics of the dominant failure mechanisms that would occur in the field for the device.

2.3.1. *Standard HTOL*

The standard high-temperature operating life (HTOL) qualification test is usually performed as the final qualification step of a semiconductor manufacturing process. The test consists of stressing some number of parts, usually about 100, for an extended time, usually 1,000 hours, at an accelerated voltage and temperature. Two features shed doubt on the accuracy of this procedure. One feature is the lack of sufficient statistical data and the second is that companies generally present zero failure results for their qualification tests and hence stress their parts under relatively low-stress levels to guarantee zero failures during qualification testing.

Unfortunately, with zero failures, no statistical data is acquired. Another feature is their calculation of the acceleration factor (AF). If the qualification test results in zero failures, this allows the assumption (with only 60% confidence!) that no more than ½ a failure occurred during the accelerated test. This would result, based on the example parameters, in a reported FIT = 5,000/AF, which can be almost any value from less than 1 FIT to more than 500 FIT, depending on the conditions and models used for the voltage and temperature acceleration.

The accepted approach for measuring FIT would be reasonably correct if there were only a single dominant failure mechanism that is excited equally by either voltage or temperature. Additionally, this same mechanism is the only one that is accelerated by the burn-in or accelerated test. For example, electromigration is known to follow Black's equation and is accelerated by increased stress current in a wire or by increased temperature of the device. If, however, multiple failure mechanisms are responsible for device failures, each failure mechanism should be modeled as an individual "element" in the system and the component survival is modeled as the survival probability of all the "elements" as a function of time (Bernstein et al. 2006).

2.3.2. *Multiple mechanisms*

If multiple failure mechanisms, instead of a single mechanism, are assumed to be time-independent and independent of each other, FIT (constant failure rate approximation) should be a reasonable approximation for realistic field failures. Under the assumption of multiple failure mechanisms, each will be accelerated differently depending on the physics that are responsible for each mechanism. If, however, an HTOL test is performed at an arbitrary voltage and temperature for acceleration based only on a single failure mechanism, then only that mechanism will be accelerated. In that instance, which is generally true for most devices, the reported FIT (especially the one based on zero failures) will be meaningless with respect to other failure mechanisms.

While the failure rate qualification has not improved over the years, the semiconductor industry understanding of reliability physics of semiconductor devices has advanced enormously. Every known failure mechanism is so well understood and the processes are so tightly controlled that electronic components are designed to perform with reasonable life and with no single dominant failure mechanism. Standard HTOL tests generally reveal multiple failure mechanisms during testing, which would suggest also that no single failure mechanism would dominate the FIT rate in the field. Therefore, in order to make a more accurate model for FIT, a preferable approximation should be that all failures are equally likely and the resulting overall failure distribution resembles a constant failure rate process that is consistent with the Mil-handbook, FIT rate approach.

The acceleration of a single failure mechanism FM_i is a highly nonlinear function of temperature and/or voltage. The temperature acceleration factor (AF_{i_T}) and voltage acceleration factor (AF_{i_V}) can be calculated separately and are the subject of most studies of reliability physics. The total acceleration factor of a failure mechanism AF_{FMi} due to different stress combinations is the product of the acceleration factors of temperature and voltage (in the example as per an exponential law, but could be a power law or a linear law),

$$AF_{FM_i} = \frac{\lambda_{FM_i}(T_{ref}, V_{ref})}{\lambda_{FM_i}(T_{use}, V_{use})} = AF_{i_T} \cdot AF_{i_V} = e^{\left[\frac{Ea}{k}\cdot\left(\frac{1}{T_{use}} - \frac{1}{T_{ref}}\right)\right]} \cdot e^{[\gamma_1 \cdot (V_{ref} - V_{use})]} \qquad [2.24]$$

This acceleration factor model is widely used as the industry standard for device qualification. However, it only approximates a single dielectric breakdown type of failure mechanism and does not correctly predict the acceleration of other mechanisms.

To be even approximately accurate, however, electronic devices should be considered to have several failure modes degrading simultaneously. Each mechanism "competes" with the others to cause an eventual failure. When more than one mechanism exists in a system, then the relative acceleration of each one must be defined and averaged at the applied condition. Every potential failure mechanism should be identified and its unique AF_{FMi} should then be calculated for each mechanism at a given temperature and voltage, so that the FIT rate can be approximated for each mechanism separately. Then, the final FIT will be the sum of the failure rates per mechanism, as described by:

$$FIT_{total} = FIT_1 + FIT_2 + \ldots + FIT_i \qquad [2.25]$$

where each i mechanism leads to an expected failure unit per mechanism, FIT_i. Unfortunately, again, individual failure mechanisms are not uniformly accelerated by a standard HTOL test, and the manufacturer is forced to model a single acceleration factor that cannot be combined with the known physics of failure models.

2.3.3. *Acceleration factor*

The qualification of device reliability, as reported by a FIT rate, must be based on an acceleration factor, which represents the failure model for the tested device. If we assume that there is no failure analysis (FA) of the devices after the HTOL test, or that the manufacturer will not report FA results to the customer, then a model should be made for the acceleration factor, AF, based on a combination of competing mechanisms. This will be explained by way of example. Suppose there are two identifiable, constant rate competing failure modes (assume an exponential distribution). One failure mode is accelerated only by temperature. We denote its failure rate as $\lambda_{FM\_1}(T)$. The other failure mode is only accelerated by voltage, and the corresponding failure rate is denoted as $\lambda_{FM\_2}(V)$.

By performing the acceleration tests for temperature and voltage separately, we can get the failure rates of both failure modes at their corresponding stress conditions. Then, we can calculate the acceleration factor of the mechanisms FM_i (for i = 1 and 2). If for the first failure mechanism $i=1$, we have $\lambda_{FM\_1}(T_{use})$, $\lambda_{FM\_1}(T_{ref})$, and for the second failure mechanism $i=2$, we have $\lambda_{FM\_2}(V_{use})$, $\lambda_{FM\_2}(V_{ref})$, then the temperature acceleration factor for the temperature at any given V is:

$$AF_{1_T}(V) = \frac{\lambda_{FM_1}(T_{use},V)}{\lambda_{FM_1}(T_{ref},V)} \quad \text{and} \quad AF_{2_T}(V) = \frac{\lambda_{FM_2}(T_{use},V)}{\lambda_{FM_2}(T_{ref},V)} \quad \text{Tuse < Tref} \qquad [2.26]$$

and the voltage acceleration factor at any given T is:

$$AF_{1_V}(T) = \frac{\lambda_{FM_{use}}(T,V_{use})}{\lambda_{FM_{ref}}(T,V_{ref})}, \quad \text{and} \quad AF_{2_V}(T) = \frac{\lambda_{FM_{use}}(T,V_{use})}{\lambda_{FM_{ref}}(T,V_{ref})} \quad \text{Vuse < Vref} \qquad [2.27]$$

The system acceleration factor between the stress ref and use conditions of (T_{ref}, V_{ref}) and (T_{use}, V_{use}) is:

$$AF = \frac{\lambda_{FM_1}(T_{use},V_{ref})+\lambda_{FM_2}(T_{ref},V_{use})}{\lambda_{FM_1}(T_{ref},V_{ref})+\lambda_{FM_2}(T_{ref},V_{ref})} \qquad [2.28]$$

The above equation can be transformed into the following two expressions:

$$AF = \frac{AF_{1T}(V)\cdot\lambda_{FM_1}(T_{ref},V_{ref})+AF_{1V}(T)\cdot\lambda_{FM_2}(T_{ref},V_{ref})}{\lambda_{FM_1}(T_{ref},V_{ref})+\lambda_{FM_2}(T_{ref},V_{ref})} \qquad [2.29a]$$

or

$$AF = \frac{\lambda_{FM_1}(T_{use},V_{use})+\lambda_{FM_2}(T_{use},V_{use})}{\frac{\lambda_{FM_1}(T_{use},V_{use})}{AF_{1T}(V)}+\frac{\lambda_{FM_2}(T_{use},V_{use})}{AF_{1V}(T)}} \qquad [2.29b]$$

These two equations can be simplified based on different assumptions.

When $\lambda_1(T_{ref}, V_{ref}) = \lambda_2(T_{ref}, V_{ref})$ (i.e. equal probability at-ref condition), equation [2.29b] gives:

$$AF = \frac{AF_{1T}(V)+AF_{1V}(T)}{2} \qquad [2.30]$$

Therefore, unless the temperature and voltage are carefully chosen so that AF_T and AF_V are very close, within a factor of about 2, then one acceleration factor will overwhelm the failures at the accelerated conditions. Similarly, when $\lambda_1(T_{use}, V_{use}) = \lambda_2(T_{use}, V_{use})$, then AF will take this form:

$$AF = \frac{2}{\frac{1}{AF_{1T}(V)} + \frac{1}{AF_{1V}(T)}} \quad [2.31]$$

and the acceleration factor applied to at-use conditions will be dominated by the individual factor with the smallest acceleration. In either situation, the accelerated test does not accurately reflect the correct proportion of acceleration factors based on the understood physics of failure mechanisms.

This discussion can be generalized to incorporate situations with more than two failure modes. Suppose a device has n independent failure mechanisms, and λ_{i_ref} represents the *ith* failure mode at reference condition, λ_{i_use} represents the *ith* failure mode at use condition, then the total acceleration factor AF can be expressed. If the device is designed so that the failure modes have equal frequency of occurrence during the use conditions:

$$\left.\begin{array}{l}
\lambda_i(S_{1use}, S_{2use}, \dots, S_{juse}) = \lambda_i\left(S_{1ref}, S_{2ref}, \dots, S_{jref}\right) \\
F_{iS_{j\neq 1}}(S_1) \qquad \text{for } S_j = S_1 \text{ to } S_p \text{ stressors} \\
\\
\lambda_i(S_{1use}, S_{2use}, \dots, S_{juse}) = \lambda_i\left(S_{1ref}, S_{2ref}, \dots, S_{jref}\right) \\
AF_{iS_{j\neq 2}}(S_2) \qquad \text{for } i = 1 \text{ to } nth \text{ failure mechanism}
\end{array}\right\} \quad [2.32]$$

The system acceleration factor between the stress ref and use conditions of $(S_{1\_ref}, S_{2\_ref}, \dots, S_{p\_ref})$ and $(S_{1\_use}, S_{2\_use}, \dots, S_{p\_use})$ reduces to:

$$AF = \frac{\lambda_1(S_{1use}, S_{2use}, \dots, S_{juse}) + \lambda_2(S_{1use}, S_{2use}, \dots, S_{juse}) + \dots + \lambda_n(S_{1use}, S_{2use}, \dots, S_{puse})}{\frac{\lambda_1(S_{1use}, S_{2use}, \dots, S_{juse})}{AF_{iS_{j\neq 1}}(S_1))} + \frac{\lambda_2(S_{1use}, S_{2use}, \dots, S_{juse})}{AF_{iS_{j\neq 2}}(S_2)} + \dots + \frac{\lambda_2(S_{1use}, S_{2use}, \dots, S_{juse})}{AF_{iS_{j\neq p}}(S_p)}} \quad [2.33]$$

For $\lambda_1(S_{1use}, S_{2use}, \dots, S_{juse}) = \lambda_2(S_{1use}, S_{2use}, \dots, S_{juse}) = \dots = \lambda_n(S_{1use}, S_{2use}, \dots, S_{puse}) = \lambda(S_{1use}, S_{2use}, \dots, S_{puse})$, the AF factor becomes:

$$AF = \frac{n \cdot \lambda(S_{1use}, S_{2use}, \dots, S_{juse})}{\lambda(S_{1use}, S_{2use}, \dots, S_{juse}) \cdot \left[\frac{1}{AF_{iS_{j\neq 1}}(S_1))} + \frac{1}{AF_{iS_{j\neq 2}}(S_2)} + \dots + \frac{1}{AF_{iS_{j\neq p}}(S_p)}\right]} = \frac{n}{\sum_{i=1}^{n} \frac{1}{AF_{iS_{j\neq i}}(S_i)}} \quad [2.34]$$

From these relations, it is clear that only when acceleration factors for each mode are almost equal, that is, $AF_1 \approx AF_2$, does the total acceleration factor become $AF = AF_1 = AF_2$, and certainly not the product of the two (as in the current model used by industry). If, however, the acceleration of one failure mode is much greater than the second, the standard FIT calculation (**[2.24]** and **[2.25]**) could be incorrect by many orders of magnitude.

Due to the exponential nature of acceleration factor as a function of V or T, if only a single parameter is changed, then it is not likely for more than one mechanism to be accelerated significantly compared to the others for any given V and T. As we will see in the next section, at least four mechanisms should be considered. Also, the various voltage and temperature dependencies must be considered in order to make a reasonable reliability model for electron devices. Until now, the assumption of equal failure probability at-use conditions has been used since it is the most conservative approach, assuming the correct proportionality cannot be determined.

2.3.4. *Proportionality matrix solution*

The basic method for solving the system of equations is described by Bernstein (White and Bernstein 2008) using the suggestion of a sum-of-failure-rate method, as described in the JEDEC Standard JEP122G, published more recently (Bernstein 2014). The matrix method forms the basis for this work. It is clear that the manufacturers of electronic components recognize the importance of combining failure mechanisms in a sum-of-failure-rate method. Also, the formula for each mechanism is well studied and documented.

We use the matrix approach to model useful life failure rate (FIT) for components in electronic assemblies by assuming each component is composed of multiple sub-components, for example; a certain percentage is effectively ring-oscillator, static SRAM, DRAM, etc. Each type of circuit, based on its operation, can be seen to affect the potential steady-state (defect related) failure mechanisms differently based on the accelerated environment, for example: electromigration, hot-carrier, NBTI, etc. Each mechanism is known to have its own acceleration factor with voltage, temperature, frequency, cycles, etc. Each sub-component will be modeled to approximate the relative likelihood of each mechanism per sub-component. Then, each component can be seen as a matrix of sub-components, each with its own relative weight for each possible mechanism.

Hence, the standard system reliability FIT can be modeled using traditional 217^+-type of algorithms and adapted to known system reliability tools; however, instead of treating each component as individuals, we propose treating each complex component as a series system of sub-components, each with its own reliability matrix.

The prediction of system reliability can be described using a linear matrix solution. Although until today, the methodology was consolidated on the microelectronic device failure mechanism, it applies directly to additional mechanisms, including thermal and mechanical stresses due to wafer bonding, and on any failure mechanism that can be modeled by the physics of failure, including wide-band-gap semiconductors and even packaging failures.

While each intrinsic mechanism is known to have different statistical distributions, the combination of distributions becomes, at the ensemble level, an approximately constant rate, as demonstrated by Drenick (1960). In its theorem, Drenick suggests and justifies the summation of the failure rate approach, which is also explained in the JEDEC handbook.

The mechanism matrix is described in Table 2.2. Each row of the matrix describes various operating conditions under which the system is tested. Each experiment, *i*, is operated with its unique voltage, frequency and temperature. The "results" column, $FIT_{i,}$, is the average time when the failure occurs under the experimental condition, which is associated with a pre-determined failure point. The example studied uses 10% performance degradation as the failure point; however, any reasonable value will work as long as it is consistent with the application. The result FIT_i is a failure rate (λ) and measured as $10^9/MTTF$.

	HCI	**BTI**	**EM**	**Results**
$\mathbf{V_1, f_1, T_1}$	$X{\cdot}A_1$	$Y{\cdot}B_1$	$Z{\cdot}C_1$	FIT_1
$\mathbf{V_2, f_2, T_2}$	$X{\cdot}A_2$	$Y{\cdot}B_2$	$Z{\cdot}C_2$	FIT_2
$\mathbf{V_3, f_3, T_3}$	$X{\cdot}A_3$	$Y{\cdot}B_3$	$Z{\cdot}C_3$	FIT_3

Table 2.2. *M-TOL matrix used to solve models with measured times to fail (Bernstein et al. 2014)*

Let us assume that each mechanism (A–C) affects the system linearly with its own acceleration factor (AF) for a given frequency. The acceleration factor formulas are shown in Table 2.3. Each equation is calculated with the experimental condition of each result on the right-hand side.

Hot carrier injection	$A_i \equiv AF_{HCI} = \frac{f}{f_0} \cdot \left(\frac{V}{V_0}\right)^N \cdot e^{-\frac{Ea_{HCI}}{k}\left(\frac{1}{T_0} - \frac{1}{T_{ac}}\right)}$
Negative bias temperature instability	$B_i \equiv AF_{NBTI} = e^{\gamma_{BTI} \cdot (V_G - V_{G,0})} \cdot e^{-\frac{Ea_{BTI}}{k}\left(\frac{1}{T_0} - \frac{1}{T_{ac}}\right)}$
Electromigration	$C_i \equiv AF_{EM} = \frac{f}{f_0} \cdot \left(\frac{V}{V_0}\right)^M \cdot e^{-\frac{Ea_{EM}}{k}\left(\frac{1}{T_0} - \frac{1}{T_{ac}}\right)}$

Table 2.3. *Equations for the acceleration factor matrix (Bernstein et al. 2014)*

Then, the matrix is solved to find a set of constants, P_i, shown here as *X–Z*, across the whole matrix, which matches the experimental results with the calculated acceleration factors. This linear matrix is solved by multiplying the inverse matrix, AF^{-1}, with lambda at each condition, as shown in Table 2.4. The solution gives the coefficients (X–Z), which make up the relative contribution of each failure mechanism on the system.

$$\overbrace{\begin{bmatrix} A_1 & B_1 & C_1 \\ A_2 & B_2 & C_2 \\ A_3 & B_3 & C_3 \end{bmatrix}}^{AF} \cdot \overbrace{\begin{bmatrix} X \\ Y \\ Z \end{bmatrix}}^{P_i} = \overbrace{\begin{bmatrix} \lambda_1 \\ \lambda_2 \\ \lambda_3 \end{bmatrix}}^{\lambda}$$

$$(AF) \cdot (Pi) = (\lambda) \rightarrow (Pi) = (AF)^{-1} \cdot (\lambda)$$

Table 2.4. *Matrix solution (Bernstein et al. 2014)*

In this matrix, the three left-hand columns describe the voltage, temperature and frequency used for the accelerated test. The normalized relative acceleration factor based on the expression for each failure mechanism is given in each of the three columns that follow. The two right-hand columns show the measured, extrapolated failure rate and the calculation based on the matrix solution. These values are extrapolated failure rates of the product after the accelerated test. This matrix will have a unique solution that will fit the percentages of each mechanism (P_i) with the measured failure rate, FIT. We found no influence of TDDB for these experiments, so we did not include them in the matrix; we only had three equations solved with three measured FIT rates.

An absolute FIT value is determined in the next row based on the mean time to fail. This allows the calibration of the final results in operation. The column line is the expected FIT (failures per billion part-hours) at those conditions. By substituting these percentages into the matrix, the true acceleration factors are determined, not only for the tested condition, but also for any extrapolated condition.

A failure mechanism matrix can be solved at any given set of conditions, that is, voltage, temperature and frequency, as a percentage at each stressed operating condition, there is a unique proportion of each mechanism for a given set of stressed conditions that will result in the given time to fail. For example, the matrix given in Table 2.4 shows the calculated acceleration factor for each mechanism contributing to a given failure in Table 2.5. The columns are normalized to their relative calculated contribution out of 100% including: EM, HCI and NBTI. A calculated reliability curve is shown in Figure 2.26, illustrating the full range of expected FIT versus temperature for any set of operational conditions shown in Table 2.6.

Volt	T °C	Freq (MHz)	EM	HCI	NBTI	Measured FIT	Calculated FIT
2.4	-20	500	0.00%	100.00%	0.00%	8.00E+04	8.00E+04
1.2	140	500	100.00%	0.00%	0.00%	5.40E-01	5.40E-01
2.4	160	0.02	0.58%	3.40%	96.02%	1.78E-02	1.78E-02
3	0	500	0.00%	81.16%	18.84%	3.45E+07	3.43E+07
2.4	173	500	35.19%	61.50%	3.31%	1.76E+01	1.73E+01
2.4	160	500	14.56%	85.34%	0.10%	1.50E+01	1.77E+01
3	0	0.20	0.00%	0.02%	99.98%	6.37E+06	6.47E+06

Table 2.5 *Table V-5. Test results showing proportions of failure mechanisms for the given V, T and F compared with the calculated as well as the measured failure rate (FIT)*

Volt	T °C	F (MHz)	EM	HCI	NBTI	Calculated FIT
1.1	30	0.02	99.76%	0.24%	0.00%	2.87E-10
1.2	70	500	100.00%	0.00%	0.00%	9.88E-04
1.5	80	500	98.58%	1.42%	0.00%	3.00E-03
2	-50	500	0.00%	100.00%	0.00%	3.12E+03
1.8	125	500	98.23%	1.77%	0.00%	1.86E-01
2	150	0.02	96.20%	3.80%	0.00%	5.16E-05

Table 2.6 *Table V-6. Calculated FIT based on the solved matrix for typical use conditions*

The 45 nm technology shows frequency-related effects at both low temperatures (below 5°C) due to HCI and at high temperatures. It is observed that the high voltage bias (1.2 V) enhances the effect of frequency, which reduces the overall HCI contribution at low frequency. The dominant failure mechanism at medium ambient temperature (between 10°C and 150°C) is related to NBTI, while the EM failure mechanism is instead observed at high temperatures.

The unique solution that solves all three equations with the three extrapolated acceleration factors gives a percentage contribution for each of the failure mechanisms. We report the reliability as FIT, which is 10^9/MTTF for each condition. The percentages for each mechanism (BTI, HCI and EM) are plotted in Figure 2.26, based on the relative contributions that were extrapolated and combined from the physics of failure equations normalized to the measured FIT of each test.

The most important result from our study is that electromigration and BTI are the most dominant failure mechanisms throughout the useful range of device operation. This is surprising since the standard HTOL test emphasizes only TDDB and BTI, as those are most accelerated by high voltage and temperature; however, under use conditions, the other two are the most important. Furthermore, it is important to see that at a very low temperature and high frequency, HCI is the most important failure mechanism and this could have very important implications for satellite and low-temperature military applications. Fortunately, very low FIT values are found, and reliability is confidently predicted.

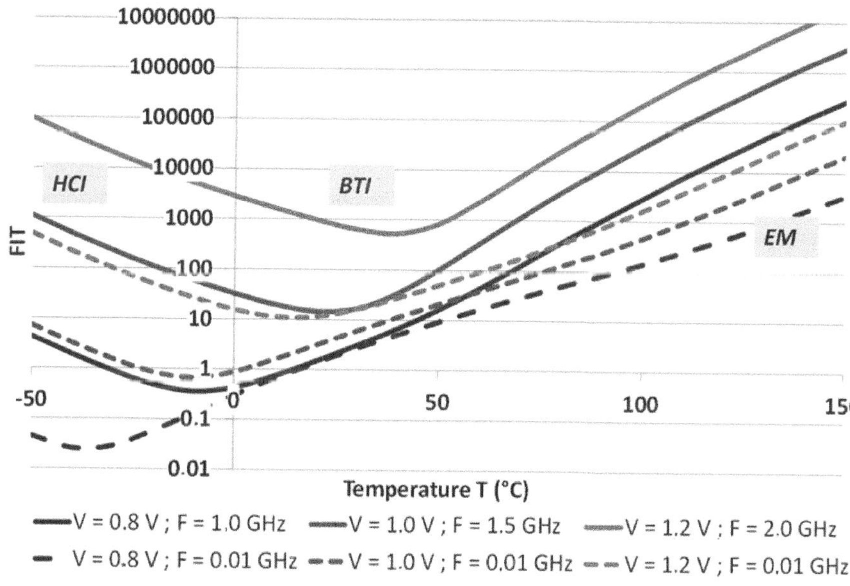

Figure 2.26. *Example of reliability curves for 45 nm DSM technology showing FIT versus temperature for voltages above and below nominal (1.2 V) and frequencies from 10 MHz (dashed line) to 2 GHz (solid line) (Bensoussan 2017). For a color version of this figure, see www.iste.co.uk/elhami/mechatronics.zip*

We present here a simple and accurate way to combine the physics of failure equations for reliability prediction from accelerated life testing. We show that a matrix approach allows the reliability physics equations to fit proportionally to the results of monitored accelerated life testing in order to extrapolate the failure rate that would be expected given actual operating parameters. This methodology can be extended to include radiation effects, frequency and even packaging and solder joint effects to give a complete system reliability evaluation framework. This matrix provides a very cost-effective way to predict reliability based on the physics of failure using only three tests, compared to the normal single-mechanism approach.

2.4. Conclusion

In this chapter, we have presented how the basic concepts developed can be applied to high reliability applications when using COTS devices or new emerging technologies in harsh environmental conditions.

A short state of the art was developed, which presented an overview of DSM and GaN technologies on deep-scale integrated technology down to 3 nm and the latest developments on GaN on Si power switch technologies.

Multiple failure mechanisms and physics of degradation in semiconductors may occur in a single set of Physics-of-Healthy time-to-failure data, and be associated with thermodynamics modeling and mathematical approaches.

Some case studies were also highlighted for applications within the automotive and aerospace industries. It affords a series of numerical application supported by experiments on a FinFET technology assuming HCI, BTI and EM as major failure mechanisms. We have shown how activation energy is related to the stress and temperature applied. It is obvious that activation energy can no longer be considered constant to extrapolate some experiments under high stress to the nominal mission operation profile.

We have also examined and shown that the equivalent activation energy increases with temperature and decreases with an increase of electrical stress conditions (named stressors).

We proposed some exercises based on the PFM level 3 for automotive applications, starting from component selection, performing technology analysis sometimes supported by constructional analysis, establishing failure rate models identified and adjusted from the literature and provided by the manufacturer.

In relation to aerospace and automotive electronic reliability and the practical application of MTOL based on the Bernstein approach, the prediction of system reliability has been described using a linear matrix solution.

2.5. References

Alam, M. (2009). 8. NBTI time dependence: Stress phase. Purdue University Course, West Lafayette, IN.

Alam, M., Kufluoglu, H., Varghese, D., Mahapatra, S. (2007). A comprehensive model for PMOS NBTI degradation: Recent progress. *Microelectronics Reliability*, 47, 853–862.

Bechtold, L., Redman, D., Tawfellows, B. (2014). Semiconductor reliability using random and wearout failure models. *Reliability and Maintainability Symposium (RAMS)*, Colarado Springs, CO.

BELLÇORE (2011). SR-332, Reliability Prediction Procedure for Electronic Equipment. BELLCORE Telcordia.

Bensoussan, A. (2017). Microelectronics reliability models for more than moore nanotechnology products. *FACTA UNIVERSITATIS, Series: Electronics and Energetics*, 30(11), 1–25, March.

Bensoussan, A. and Suhir, E. (2013). Design-for-reliability (DfR) of aerospace electronics: Attributes and challenges. *IEEE Aerospace Conference*, 2–9 March.

Bensoussan, A., Durier, A., Bernstein, J.B. (2017). M-Storm: A new approach to model electronics components reliability. *Automotive Power Electronics (APE 2017)*, Paris.

Bernstein, J.B. (2014). *Reliability Prediction from Burn-In Data Fit to Reliability Models*. Elsevier, London.

Bernstein, J.B. (2015). Reliability prediction for aerospace electronics. *IEEE Aerospace Conference*, Big Sky, MN.

Bernstein, J.B., Gurfinkel, M., Li, X., Walters, J., Shapira, Y., Talmor, M. (2006). Electronic circuits reliability modelling. *Microelectronics Reliability*, 46, 1957–1979.

Bernstein, J.B., Gabbay, M., Delly, O. (2014). Reliability matrix solution to multiple mechanism prediction. *Microelectronics Reliability Journal*, 54, 2951–2955.

Blonkowski, S. (2010). Filamentary model of dielectric breakdown. *Journal of Applied Physics*, 107.

Bravaix, A. (2019a). Part 1 – FRAME Analysis IRT-ISEN. Internal report, IRT-ISEN, Toulouse, September.

Bravaix, A. (2019b). Part 2 – DDR3 FF20 Reliability Analysis – ISEN. Internal report, IRT-ISEN, Toulouse, September.

Bravaix, A. (2019c). Part 3 – FPGA FF16 Reliability Analysis – ISEN. Internal report, IRT-ISEN, Toulouse, September.

Bravaix, A., Guerin, C., Huard, V., Roy, D., Roux, J., Vincent, E. (2009). Hot-carrier acceleration factors for low power management in DC-AC stressed 40nm NMOS node at high temperature. *IEEE International Reliability Physics Symposium (IRPS) Proc.*, Anaheim, CA.

Bravaix, A., Guérin, C., Goguenheim, D., Huard, V., Roy, D., Besset, C., Renard, S., Mamy Randriamihaj, Y. (2010). Off state incorporation into the 3 energy mode device lifetime modeling for advanced 40nm CMOS node. *IRPS*, San Jose, CA.

Bravaix, A., Huard, V., Cacho, F., Federspiel, X., Roy, D. (2015). Hot-carrier degradation in decananometer CMOS nodes: From an energy-driven to a unified current degradation modeling by a multiple-carrier degradation process. In *Hot Carrier Degradation in Semiconductor Devices*, Grasser, T. (ed.). Springer, Cham.

Chang, K.-C., Liao, J.-C., Chang, T.-C., Yeh, C.-H., Lin, C.-Y., Jin, F.-Y., Lin, Y.-S., Ciou, F.-M., Hung, W.-C., Lin, Y.-H. et al. (2019). Abnormal relationship between hot carrier stress degradation and body current in high-k metal gate in the 14-nm node. *IEEE Electron Device Letters*, 40(4), 498–501.

Condra, L., Horan, G., Scofield, B. (2005). As semiconductor devices shrink so do their reliability and lifetimes. *National Software and Airborne Electronic Hardware. Standardization Conference*, Denver, CO.

Creveling, C. (1956). Increasing the reliability of electronic equipment by the use of redundant circuits. *Proc. IRE*, 44, 509–515.

Crupi, G., Schreurs, D.M.-P., Raskin, J.-P., Caddemi, A. (2013). A comprehensive review on microwave FinFET modeling for progressing beyond the state of art. *Solid-State Electronics*, 80, 81–95.

Dhillon, B. and Rayapati, S. (1986). A method to evaluate reliability of three-state device network. *Microelectronics Reliability Journal*, 26, 525–554.

FIDES (2004). A reliability methodology for electronic systems. Guide, DGA.

GIFAS (2015). Guide for the selection of Deep Sub-Micron (DSM) for aeronautic equipment – Semiconductors lifetime – GIFAS/2015/5022. Guide, Groupement Industries Françaises Aeronautiques et Spatiales, Paris.

Grasser, T. (2015). *Hot Carrier Degradation in Semiconductor Devices*. Springer, Cham.

Grasser, T., Rott, K., Reisinger, H., Wagner, P.-J., Goes, W., Schanovsky, F., Waltl, M., Toledano-Luque, M., Kaczer, B. (2013). Advanced characterization of oxide traps: Thedynamic time-dependent defect spectroscopy. *IRPS*, April.

Guerin, C., Huard, V., Bravaix, A. (2007). The energy-driven hot-carrier degradation modes on nMOSFETs. *IEEE Transactions on Device and Materials Reliability*, 7(2), 225–235.

Hirex Engineeering (2019). HRX/RCA/0123. Report, Hirex Engineering, Toulouse.

Huard, V. (2010). Two independent components modeling for Negative Bias Temperature Instability. *IRPS*, San Jose, CA.

Huard, V., Denais, M., Parthasarathy, C. (2005). NBTI degradation: From physical mechanisms to modelling. *Microelectronics Reliability*, 46, 1–23.

JEDEC (2021). JESD85. *Methods for Calculating Failure Rates in Units of FITs*. JEDEC Publication, JEDEC Solid State Technology, Arlington, VA.

Kaczer, B., Arkhipov, V., Jurczak, M., Groeseneken, G. (2005). Negative bias temperature instability (NBTI) in SiO2 and SiON gate dielectrics understood through disorder-controlled kinetics. *Microelectron Engineering*, 80, 122–125.

Kaczer, B., Degraeve, R., Roussel, P., Groeseneken, G. (2007). Gate oxide breakdown in FET devices and circuits: From nanoscale physics to system-level reliability. *Microelectronics Reliability Journal*, 47, 559–566.

Kim, Y.K., Lee, J.S., Kim, G., Park, T., Kim, H.J., Cho, Y.P., Park, Y.J., Lee, M.J. (2019). Partial isolation type saddle-FinFET-Pi-FinFET – For Saddle FinFET Sub-30 nm DRAM Cell Transistors. *Electronics*. doi:10.3390/electronics8010008 [Online]. Available at: www.mdpi.com/journal/electronics.

Lee, H., Lee, C.-H., Park, D., Choi, Y.-K. (2005). A study of negative-bias temperature instability of SOI and body-tied FinFETs. *IEEE Electron Device Letters*, 26(5), 326–328.

Liu, C., Sagong, H.-S., Kim, H., Choo, S., Lee, H., Kim, Y. (2015). Systematical study of 14 nm FinFET Reliability: From device level stress to product HTOL. *IRPS*, Monterey, CA.

Loubet, N., Hook, T., Montanini, P., Yeung, C.-W., Kanakasabapathy, S., Guillorn, M., Yamashita, T., Zhang, J., Miao, X., Wang, J. et al. (2017). Stacked nanosheet gate-all-around transistor to enable scaling beyond FinFET – T17-5. *Symposium on VLSI Technology Digest of Technical Papers*, Kyoto.

Mahapatra, S. and Rashmi, S. (2018). On the universality of hot carrier degradation: Multiple probes, various operating regimes, and different MOSFET architectures. *IEEE Transactions on Electron Devices*, 65(8), 3088–3094.

McPherson, J.W. (1986). Stress dependent activation energy. *Proc. IEEE 24th IRPS*, 12.

McPherson, J.W. (2012). Time dependent dielectric breakdown physics – Models revisited. *Microelectronics Reliability Journal*, 52, 1753–1760.

McPherson, J.W. (2019). ESREF Tutorials 2019. The future of reliability testing – Can zero failures become reality? *ESREF, European Symposium on Reliability of Electron Devices*, Toulouse.

McPherson, J.W., Khamankar, R., Shanware, A. (2000a). Complementary model for intrinsic TDDB in SiO2 dielectrics. *Journal of Semiconductor Science and Technology*, 15, 462.

McPherson, J.W., Khamankar, R.B., Shanware, A. (2000b). A complementary molecular-model (including field and current) for TDDB in SiO2 dielectrics. *Microelectronics Reliability*, 40, 1591–1597.

Meneghini, M., Meneghesso, G., Zanoni, E. (2017). Power GaN devices: Materials, applications and reliability. In *Power Electronics and Power Systems*, Chow, J.H., Stankovic, A.M., Hill, D. (eds). Springer, Berlin.

Monsieur, F., Vincent, E., Roy, D., Bruyere, S., Vildeuil, J., Pananakakis, G., Ghibaudo, G. (2002). A thorough investigation of progressive breakdown in ultra-thin oxides. Physical understanding and application for industrial reliability assessment. *IRPS*, Dallas, TX.

Nicollian, P. (2007). Physics of trap generation and electrical breakdown in ultra-thin SiO2 and SiON gate dielectric materials. Thesis, University of Twente, Twente.

Nicollian, P., Rodder, M., Grider, D., Chen, P., Wallace, R., Hattangady, S. (1999). Low voltage stress-induced-leakage-current in ultrathin gate oxides. *37th Annual International Reliability Physics Symposium*, San Diego, CA.

Nikulin, M., Gerville-Reache, L., Couallier, V. (2007). *Statistique des essais accélérés.* Hermes Science-Lavoisier, Paris.

Nikulin, M., Couallier, V., Gerville-Reache, L. (2009). Accelerated life and degradation models in reliability and safety: An engineering perspective. *EA 2961, Mathematical Statistics and Applications*. University Victor Segalen Bordeaux 2, Bordeaux.

PRISM Now 217Plus (TM) (n.d.). RIAC – The Reliability Analysis Center [Online]. Available at: http://www.theriac.org.

QUANTERION (2015). *Handbook of 217Plus Reliability Prediction Models.* QUANTERION Solution Inc., Utica, NY.

Ramey, S., Ashutosh, A., Clifford, J., Hattendorf, M., Hicks, J., James, R., Rahman, A., Sharma, V., St Amour, A., Wiegand, C. (2013). Intrinsic transistor reliability improvements from 22nm Tri-Gate technology. *IRPS*, Anahiem, CA.

Randomson, H.H., Zhang, Y., He, X., Cui, H., Li, J., Xiang, J., Liu, J., Gu, S., Wang, G. (2007). The challenges of advanced CMOS process from 2D to 3D. *Applied Sciences*, 7, 1–32.

Rauch, S. and La Rosa, G. (2005). The energy driven paradigm of NMOSFET hot carrier effects. *IRPS*, San Jose, CA.

Regis, D., Berthon, J., Hubert, G., Gatti, M. (2014). DSM reliability concerns – Impact on safety assessment. *SAE International Journal of Aerospace*, 7(1).

Ribes, G., Bruyere, S., Roy, D., Denais, M., Roux, J., Parthasarathy, C., Huard, V., Ghibaudo, G. (2006). New extensive MVHR breakdown models. *IRPS*, San Jose, CA.

Roesch, W. (2000). Volume and quality impacts on reliability: A new game for GaAs. *GaAs REL Workshop*, Seattle, Washington.

Saliva, M., Cacho, F., Ndiaye, C., Huard, V., Angot, D., Bravaix, A., Anghel, L. (2015). Impact of gate oxide breakdown in logic gates from 28nm FDSOI CMOS technology. *IRPS*, Monterey, CA.

Serma Technologies (1996). Alcatel Espace Internal failure analysis report EL 4103. Report, Serma Technologies, Toulouse, March.

Shin, C., Kim, K., Kim, J., Ko, W., Yang, Y., Lee, S., Jun, C.S., Kim, Y.S. (2013). Fast, exact, and non-destructive diagnoses of contact failures in nano-scale semiconductor device using conductive AFM. *Scientific Reports*, 3(2088).

Singh, A.K. and Samanta, J. (2012). Different physical effects in UDSM MOSFET for delay and power estimation: A review. *IEEE Students' Conference on Electrical, Electronics and Computer Science*, Bhopal.

Srinivasan, J. and Adce, S.V. (2003). The case for microarchitectural awareness of lifetime reliability. Technical report, UIUC CS n° UIUCDCS-R-2003-2391.

Stathis, J., Mahapatra, S., Grasser, T. (2018). Controversial issues in negative bias temperature instability. *Microelectronics Reliability Journal*, 81, 244–251.

Suhir, E. (2013). Predicted reliability of aerospace electronics: Application of two advanced concepts. *IEEE Aerospace Conference*, 2–9 March.

Takeda, E. and Suzuki, N. (1983). An empirical model for device degradation due to hot-carrier injection. *IEEE Electron Device Letters*, EDL-4(4), 111–113.

Veendrick, H.J. (2017). *Nanometer CMOS ICs – From Basics to ASICs, Heeze*, 2nd edition. Springer, Heeze.

Verkest, D. (2018). IMEC Roadmap – Cadence Breakfast Bytes Blogs. Post from Paul McLellan, 7 June [Online]. Available at: https://community.cadence.com/cadence_blogs_8/b/breakfast-bytes/posts/imec2 [Accessed 28 February 2019].

VITA (2011). Physics of failure reliability predictions. ANSI/VITA 51.2-2011, June.

W.G. Components (2015). Guide for the selection of Deep Su-Micron (DSM) for aeronautic equipment – Semiconductor lifetime. Guide, GIFAS/2015/5022, Paris.

Wu, E. and Sune, J. (2005). Power-law voltage acceleration: A key element for ultra-thin gate oxide reliability. *Microelectronics Reliability Journal*, 45, 1809–1834.

Wu, E., Li, B., Stathis, J., Achanta, R., Filippi, R., McLaughlin, P. (2013). A time-dependent clustering model for non-uniform dielectric breakdown. *IEEE International Electron Devices Meeting*, 9–11 December.

PART 2

Failure and Analysis of Systems Engineering

3

Fault Tree Analysis in the Context of Systems Engineering Design Analysis

3.1. Introduction

The advent of autonomous vehicles has brought with it a dramatic increase in automotive product complexity due, for example, to the need for significantly increased in-vehicle communication between systems, in-vehicle to external communications and appreciably different vehicle system to human interactions. This leads to automotive vehicles having their own inherent safety challenges (Cui et al. 2019). It is also the case that automotive OEMs continue to be hit financially by product launch problems and significant after sales warranty claims and recalls. It is vital, therefore, that potential failures associated with both autonomous and conventional vehicles are identified during product design and countermeasures taken to either minimize or eliminate the chance of the failure occurring, or to mitigate the effect of the failure. A number of engineering tools are available to the automotive product designer that facilitate the identification of potential failure events and the development of countermeasures to such failure during product design, including failure mode and effects analysis (FMEA) and fault tree analysis (FTA). This chapter considers a model-based approach to FTA, as an integral part of the product design process, supporting the early stages of the design analysis and synthesis, as opposed to the more conventional use of FTA as a risk and reliability assessment tool for an existing system.

A review of some of the important aspects and related improvements that have been made to the FTA methodology since its introduction is considered next before looking at the detail of the model-based approach, followed by a brief review of the relationship between the approach and other engineering tools such as FMEA.

Chapter written by Felician CAMPEAN and Ed HENSHALL.

3.1.1. *Background to fault tree analysis*

Fault tree analysis (FTA) was first developed by Bell Laboratories in 1962 as a graphically based technique for safety analysis on the Minuteman weapon system (Lee et al. 1985). Since its inception, the methodology has gained widespread use across many industries. The methodology links a particular system failure top-down through Boolean gates to multiple potential failure events. It is used to improve the reliability, maintainability and safety of complex systems through top-down root cause analysis where the potential or actual causes of failure that occur, either in isolation or inclusively, are not all immediately obvious. The identification of such causes enables system design improvement. FTA is also widely used to predict the reliability of system design where the probability of the failure events that can lead to system failure is known or can be estimated.

Significant effort has been directed at improving the FTA methodology since its inception with a major focus of such work being directed towards improving the integrity of system reliability prediction using fuzzy logic (i.e. fuzzy set theory) when quantifying component failure rates (Nadjafi 2021; Yazdi and Zarei 2018). In recognition that most systems are dynamic, that is, exhibit time-dependent behavior, the original static FTA methodology has been extended to cope with dynamic systems, for example, through the introduction of additional logic gates featuring time dependency (Dugan 1992). These two improvement strategies were synthesized in studies of Fuzzy Dynamic FTA by Li et al. (2012). In another innovation directed at the analysis of dynamic systems, Kaiser et al. (2007) integrated state dependencies and the temporal order of events into what they term state/event fault trees.

Fault tree construction was developed, and is normally seen, as a top-down deductive process (Haasl 1965; Vesely 2002; Cristea and Constantinescu 2017; Kabir 2017). The methodology is often based on an informal description of the system being analyzed. Where this is the case, it is possible that causal events are included on the fault tree that could not lead to the system failure under investigation, or worse, failure events that could lead to the system failure are omitted. Thus, the deductive top-down approach may not identify and hence address all the complex interactions present in a large engineering product in an analytically rigorous manner (Wang and Ruxton 1997). With the rapid increase in system complexity, it becomes more difficult for the engineering team developing a fault tree to reflect all the important nuances of the system of interest (Gonschorek et al. 2018). This situation is exacerbated by the fact that conventional fault tree construction by its nature is very time-consuming and error prone with the size of any fault tree typically growing geometrically with the complexity of the system being analyzed (Cristea and Constantinescu 2017).

In view of the effort involved in constructing fault trees, attention has been directed to their automated generation, for example, as a part of what is termed model-based dependability assessment (MBDA) in which the dependability of a system is assessed in a quantitative manner once a design is complete. Dependability in this context comprises safety, reliability, availability and maintainability (Sharvia et al. 2016). Automated fault tree synthesis is generally conducted by a computer algorithm which interrogates a model of the system under investigation. This relies on the availability of a system model with which the algorithm can interact. Such models can be established in various ways, for example, through directed graphs (Ulerich and Poers 1988), state modeling such as finite state machine modeling (Liggesmeyer and Rothfield 1998) and decision tables (Andrews and Henry 1997). The technique of hierarchically performed hazard origin and propagation studies (HiP-HOPS; Sharvia and Papadopoulos 2014) uses a system model generated in a standard modeling package such as Matlab Simulink to which failure data is annotated to generate component-level fault trees (CFT) which are subsequently synthesized into a system-level tree. Similarly, Alshboul and Petriu (2018) extract component-level fault trees from SysML diagrams which include failure data and subsequently generate system-level fault trees by collation of the component-level fault trees. Hofig et al. (2018) linked a series of component fault trees (CFT) through input and output failure modes, effectively forming a system fault tree, to facilitate system-level analysis. One area where the automatic generation of fault trees has prospered is in systems where Boolean logic can readily be used to model system behavior such as the information flow in reactive electronic systems (Banach and Bozzano 2011).

All engineering activity including the development of fault trees is dependent on the knowledge of those engaged in the activity (Yi et al. 2017). Knowledge is widely classified into the two main categories of tacit and explicit (Nonaka 1991). Explicit knowledge is academic knowledge or "Know What", for example, engineering principles; while tacit knowledge, seldom expressed openly, is "Know How" acquired through experience and often resembles intuition (Smith 2001). Both types of knowledge are invaluable in FTA although historically the tendency has been to focus more on the use of explicit knowledge to the detriment of tacit knowledge since, by their nature, dissemination of the former is easier than the latter.

This chapter extends the function fault tree analysis (FFTA) approach of Campean and Henshall (2008) of basing the development of a fault tree on the functional model of a system that system state flow analysis provides. The original FFTA methodology has been significantly enhanced to take account of the advances that have been made to system state flow analysis enabling the development of the function fault tree (FFT) to be allied more rigorously and systematically to the conjugate system state flow diagram (SSFD) functional model. The functional model-based approach enables FTA to be conducted during the early design stage of a system by hypothesizing functionally based failure modes and events and so enhancing the strength of the analysis in the face of uncertain operating environments. The functional basis of the methodology is summarized next before considering the approach in some detail within a case study. While any necessary facets of fault tree analysis are clarified, the reader is directed elsewhere for a more detailed treatise on the methodology of FTA, for example, Vesely (2002).

3.1.2. *Functional basis of fault tree analysis*

The performance of a complex system can be considered either from the standpoint of the success or from the standpoint of failure as is recognized in the US Nuclear Regulatory Commission Fault Tree Handbook of 1981 (Vesely et al. 1981) and illustrated in Figure 3.1.

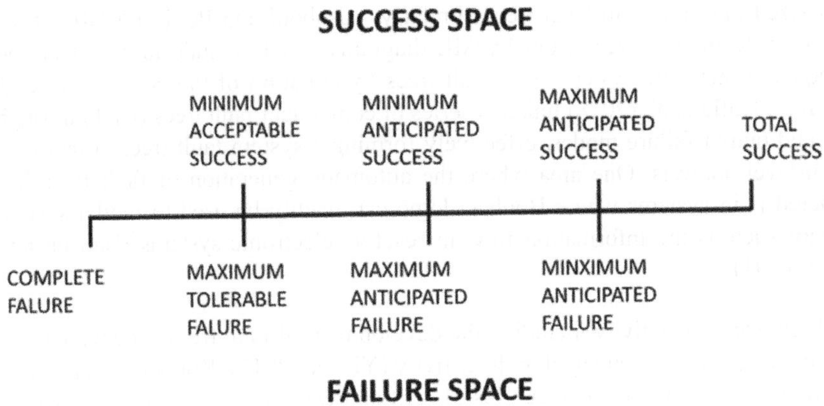

Figure 3.1. *The failure space – success space concept (Vesely et al. 1981)*

In addition to the "mirrored" relationship between success space and failure space, Figure 3.1 illustrates that there are degrees of success/failure. The model-based approach to FTA described by Campean and Henshall (2008) exploits this relationship in the context of function in the success space/domain and function failure in the failure space/domain. Function failure can be defined using the simple taxonomy described in the AIAG/VDA FMEA Handbook (2019), which is of four types: no function, partial function, intermittent function and unintended function. Partial function includes too little function, too much function and function that degrades over time; intermittent function is a function that is discontinuous, occurring some of the time but not at others. Unintended function occurs when a system functions when it is not required to function such as during a control command failure.

The approach is founded on the functional decomposition of a system based on a scientifically derived functional model of the system, namely the system state flow diagram (SSFD) (Yildrim et al. 2017), which will be introduced in the context of the case study. Basing the functional decomposition on a precise system model allows this decomposition, and any fault tree that is derived from it, to be wholly consistent with the physics of the system, so preserving the engineering integrity of any subsequent analysis of the fault tree. Establishing a fault tree through functional decomposition also allows coherent information flow between product improvement methodologies, for example, between FTA and FMEA (Henshall et al. 2014).

3.1.3. *Case study: electric bicycle drive system*

The term electric bicycle (eBike) includes bicycles that operate in different ways, for example, bicycles that:

i) can be propelled exclusively by an auxiliary motor;

ii) cannot be propelled exclusively by an auxiliary motor.

The legal use of eBikes varies globally depending upon in which of the above two categories they lie. For example, eBikes that fall into category (i) can legally be ridden in the US, but not in Europe. In Europe, eBikes within category (ii) have the advantage of being legally considered bicycles and so can be ridden without the need for a driving license. Other regional regulations apply, for example, propulsion solely with an auxiliary motor is restricted to a maximum speed, typically 32 kph (20 mph) in most US States, while pedal assistance in Europe is only available when the pedals move forward, and the assistance speed cannot exceed 25 kph (15.5 mph). We will consider the design of an eBike for use across global markets and so one that is able to fall into either category dependent on the motive control strategy that is applied. The control aspects will not be considered in the case study to ensure that the analysis does not become overly complex.

The case study follows the design of a key subsystem of an electric bicycle, namely the drive system. The design concept of a conventional bicycle with electrical power provided by a rear wheel hub motor has been selected.

3.2. System-level analysis

3.2.1. *Function analysis and decomposition*

The development of an SSFD for the eBike drive system is conducted within a system engineering framework. Analysis at system (bicycle) level includes the black box depiction of the drive system shown in Figure 3.2.

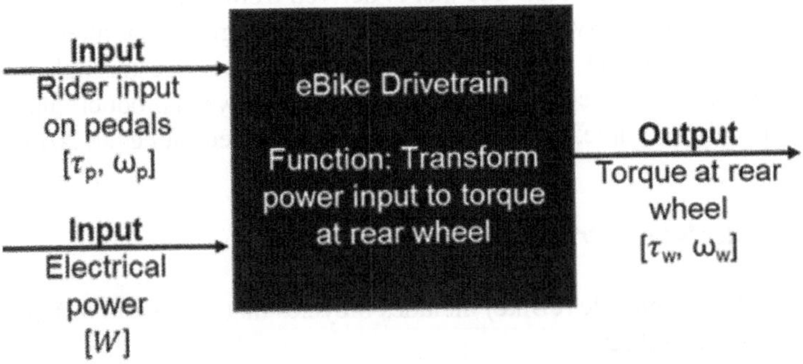

Figure 3.2. *Black box representation of eBike drive system*

In the context of an SSFD, a state is defined by Yildirim et al. (2017) as:

> an object that incorporates the "global attributes" of "time" and "space" with the attribute of time being implied by the relative position of the State in terms of State Flow and the attribute of space considered as the "location" of the object.

With an object defined by Dori (2002) as: "an object is a thing that has the potential of stable, unconditional physical or mental existence.

Based on the taxonomy of Pahl et al. (2007), a state is an object that can be material, energy or information and is depicted graphically in Figure 3.3.

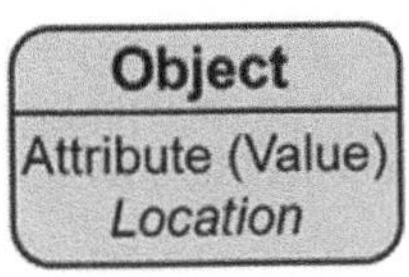

Figure 3.3. *Generic depiction of state (Yildirim 2017)*

The SSFD is a graphical model of the flow of material, energy and information that are required in a system to convert system inputs to system outputs. Thus, an eBike drive system SSFD represents graphically what happens inside the black box in Figure 3.2.

Yidrim et al. (2017) introduced a series of guidelines to facilitate the development of an SSFD, termed SSFD heuristics. The SSFD heuristics are as follows:

– Main flow heuristic: the flow of states from input(s) to output(s) that enable a system to achieve its overall function, that is, the flow that allows a system to achieve its design intent. In developing the SSFD, this is the first flow type to identify.

– Connecting flow heuristic: flows directed into the main flow to facilitate particular conversions within the main flow.

– Branching flow heuristic: flows that branch out of the main flow.

– Conditional fork node heuristic: two or more flows combining to a single flow under defined conditions, or a single flow diverging into two or more flows under defined conditions.

Each heuristic will be illustrated in the context of the case study. The heuristics facilitate the development of an SSFD by indicating what types of flow structures may be present. The structure and content of the SSFD is determined by the way in which material, energy and information flow through a system to ensure that the system behaves in a manner consistent with its design intent. Since the flow of material, energy and information is governed by the laws of physics, it is the SSFD's compliance with these laws, and not the heuristics, that determines its structure and content.

The flow between two objects/states is dependent on an action which can be defined as a function, with the entity that will generate the function being termed a design solution, that is, that which a product designer chooses or designs to generate the function. The relationship between input and output objects, the design solution and the function necessary for state flow is depicted in an object, function and design-solution triad (OFD), as shown in Figure 3.5 with the state flow being the flow of the left-hand object/state in the figure to become the right-hand object/state.

The high-level black box depiction of the eBike drive system of Figure 3.2 is represented as two OFDs: one for the human drive aspect of the eBike drive system and one for the electric drive aspect as shown in Figures 3.5a and 3.5b respectively.

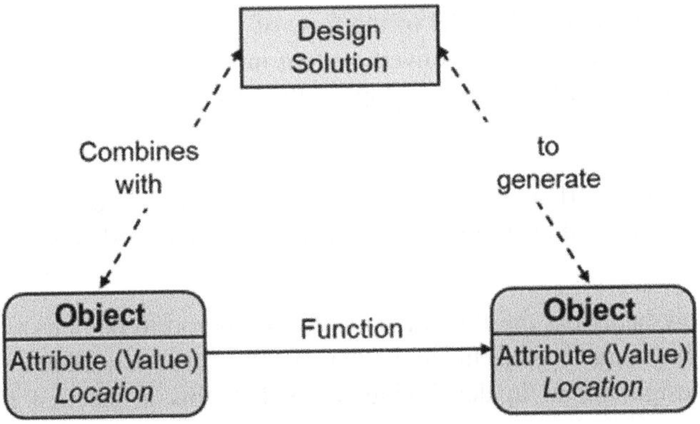

Figure 3.4. *Generic depiction of object, function and design solution triad (Yidrim 2017). For a color version of this figure, see www.iste.co.uk/elhami/mechatronics.zip*

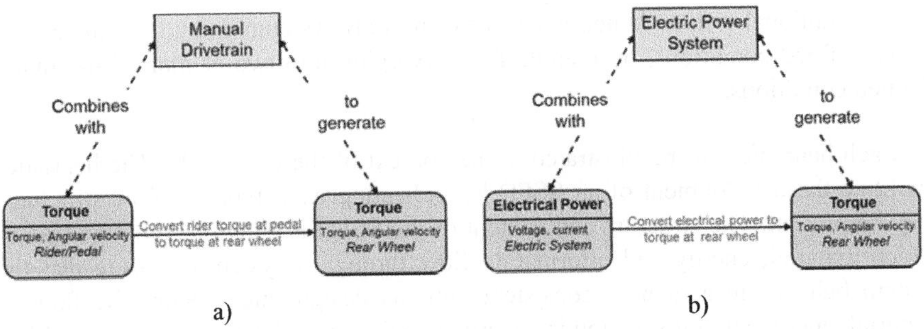

Figure 3.5. *a) OFD for the human drive system. b) OFD for the electric drive system. For a color version of this figure, see www.iste.co.uk/elhami/mechatronics.zip*

By applying the conditional fork node heuristic, the two OFDs can be combined into a single graphic to become the high-level, that is, the system level, SSFD for the eBike drive system (Figure 3.6). The fork node indicates that the eBike can operate in the two modes of pedal only (PO) and pedal assist (PA). The conditional nature of the fork is that the Rider/User can select which mode is intended. Detail on how this intent is transmitted to the system, that is, how each mode is selected, will become apparent with the next level of analysis. As can be seen in this case, the two separate state flows from the rider and electrical power inputs connect to generate the common state output of torque at the rear wheel, that is, the torque at the rear wheel is the combination of the torque supplied by the human drive system and the torque supplied by the electric drive system.

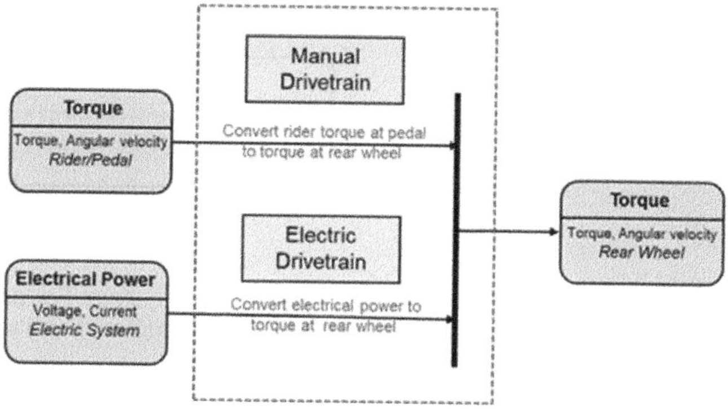

Figure 3.6. *System-level SSFD for eBike drive system. For a color version of this figure, see www.iste.co.uk/elhami/mechatronics.zip*

The impact of the SSFD flow heuristics on the SSFD structure is indicated in Figure 3.7 by the color coding in the accompanying key with two main flows and one conditional fork node.

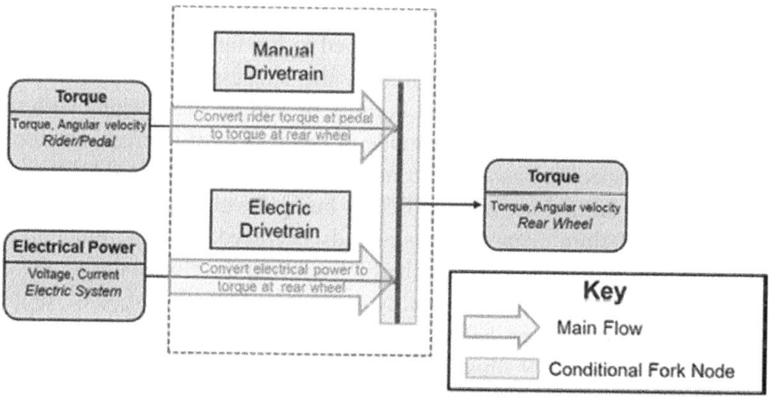

Figure 3.7. *Application of SSFD heuristics in developing system-level SSFD. For a color version of this figure, see www.iste.co.uk/elhami/mechatronics.zip*

The functional information and its structural relationship can be derived from the SSFD in the form of a function tree. The influence of the SSFD main flow heuristic and conditional fork node heuristic on the structure of the SSFD is carried across to the structure of the function tree as shown in the system-level function tree for the eBike drive system of Figure 3.8 with top most function being that of the overall system depicted in Figure 3.2.

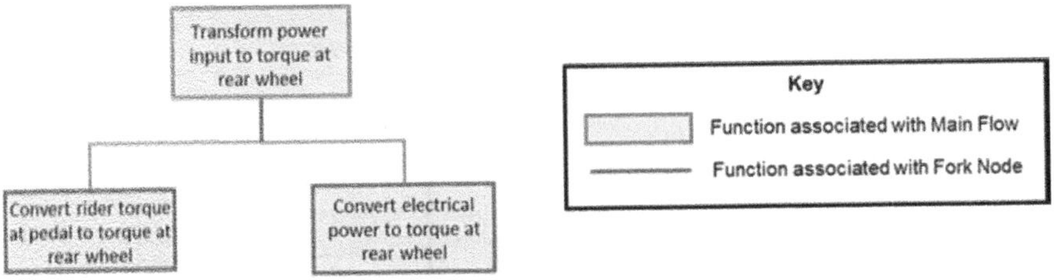

Figure 3.8. *System-level function tree for eBike drive system. For a color version of this figure, see www.iste.co.uk/elhami/mechatronics.zip*

The arrows in the SSFD of Figure 3.7 signify state flow while the lines in the function graphic of Figure 3.8 denote a structural relationship between functions. The two subfunctions in Figure 3.8 are associated both with the main flow of the system and the fork node which links them to the top function. This duality of association is depicted in the graphic by the use of dual color coding.

3.2.2. *System-level function fault tree development*

A function fault tree (FFT) can be developed in the failure domain as the mirror image of the corresponding function tree in the success domain. The FFT uses widely used fault tree terminology and definitions and as such begins with the *top event* that represents a function failure mode of the system of interest. The top event needs to be clearly specified as a failure mode based on the standard taxonomy (i.e. total failure, partial failure, etc.). If the FFT is being developed in order to facilitate the discovery of a failure event that has already occurred, then the function failure mode type will be known. However, if the FFT is being developed during the design of a system to help ensure its dependability, a separate FFT should be developed for each of the feasible and probable top-event functional failure types that can potentially occur. Figure 3.9 illustrates an eBike system-level FFT that represents the mirror image of the system-level function tree shown in Figure 3.8 for a total failure top event and total failure sub events, termed *intermediate events*.

The intermediate events are joined by an *AND* logic gate since in this case both intermediate event function failure modes must occur for the top event to occur. Note that if the human drive system power drive fails while the eBike is in PA mode, the electric powered assistance may need to be discontinued, for example, to ensure European legal regulations are not contravened. The corresponding system functions are linked to the top-most function on the function tree by an implied *OR* logic gate in which one system function *or* the other *or* both must be generated for the top most function to occur, again with the proviso, for example, that European legal regulations must be met. It is not normal practice to include logic gates symbols in a function tree since, as will be seen, the gates are invariably AND gates; this example is an exception.

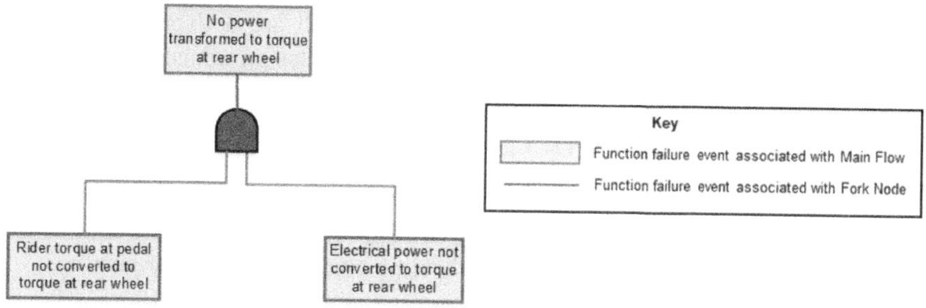

Figure 3.9. *eBike system-level function tree with total failure mode events. For a color version of this figure, see www.iste.co.uk/elhami/mechatronics.zip*

In mirroring the structure of the system-level eBike function tree, the FFT structure results from application of the main flow and conditional fork node heuristics in the development of the system-level eBike drive system SSFD from which the function tree and hence FFT are derived. This structure is due to the convergence of the two main flows in the SSFD to enable the electrically powered torque to sum in a homogeneous manner with the human torque to produce pedal assistance to the rider.

Figure 3.10 illustrates a system-level eBike drive system FFT with partial function failure mode events compared to the total failure events of Figure 3.10 with the word *reduced* reflecting a loss of efficiency such that the power output is less than that expected for given power inputs. The quantification of this term is considered later during the analysis of the FFT.

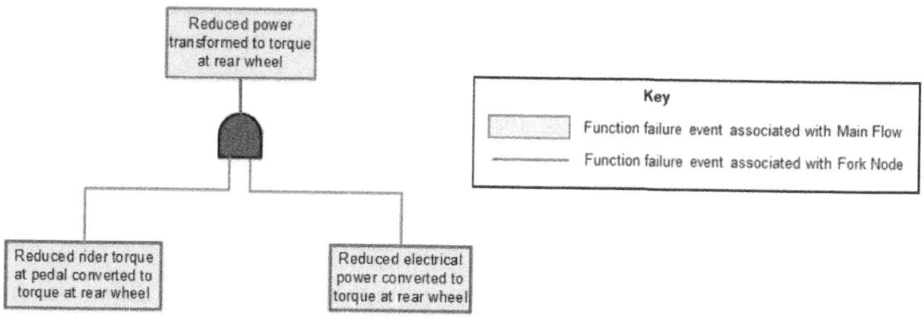

Figure 3.10. *eBike system-level function fault tree with partial function failure mode events. For a color version of this figure, see www.iste.co.uk/elhami/mechatronics.zip*

The partial function failure modes in the FFT could reflect an adverse system interaction between the two main flows through the eBike transmission and, as such, is an important design consideration pointing to the need to decouple the flows. The two main flows would be coupled if, for example, both are delivered to the rear wheel through a common transmission. The following design concept was chosen for the eBike drive system system as a result of system-level analysis:

– A brushless DC motor (BLDC) to deliver the electric powered torque to the rear wheel by direct drive at the wheel hub.

– A convention pedal and crank set to deliver the human generated torque to the rear wheel through a linked chain and freewheeling cog using derailleur gearing.

The selection of this design concept is considered further in the section on FFT analysis. Subsystem-level analysis based on this design concept is considered next.

3.3. Subsystem-level analysis

3.3.1. *Subsystem-level function decomposition*

The drive system including the conventional human drive system and gearing, and the motor and its controls are a subsystem of the eBike. The SSFD at this level is developed in the same manner as the system-level SSFD and represents an increased level of granularity over the system-level SSFD. The subsystem SSFD for the eBike drive system is shown in Figure 3.11.

The eBike drive system subsystem SSFD shows the conversion of the rider input torque at the pedal to the output torque at the rear wheel and the conversion of electrical power to output torque at the rear wheel. These conversions were represented in the system-level SSFD of Figure 3.7 as *main flows* which combine through a *conditional fork node*; the influence of these heuristics is maintained in the subsystem SSFD (see Figure 3.12). Figure 3.12 also shows the control of the motor enabling the rider to switch the electric motor on/off and select the level of pedal assistance required as a *connecting flow*. Similarly, the control of the derailleur gearing by the rider is through a *connecting flow*.

Note that the SSFD is design solution neutral, for example, the connecting flow associated with the derailleur gear change in which the rider applies a force at the *Rider Gear Selector* interface which is directed via the fork node to either the front derailleur and/or the rear derailleur is consistent with a conventional design of separate levers and cables in which front and rear gears can be changed independently or a coupled arrangement in which selected front and rear gears are changed simultaneously.

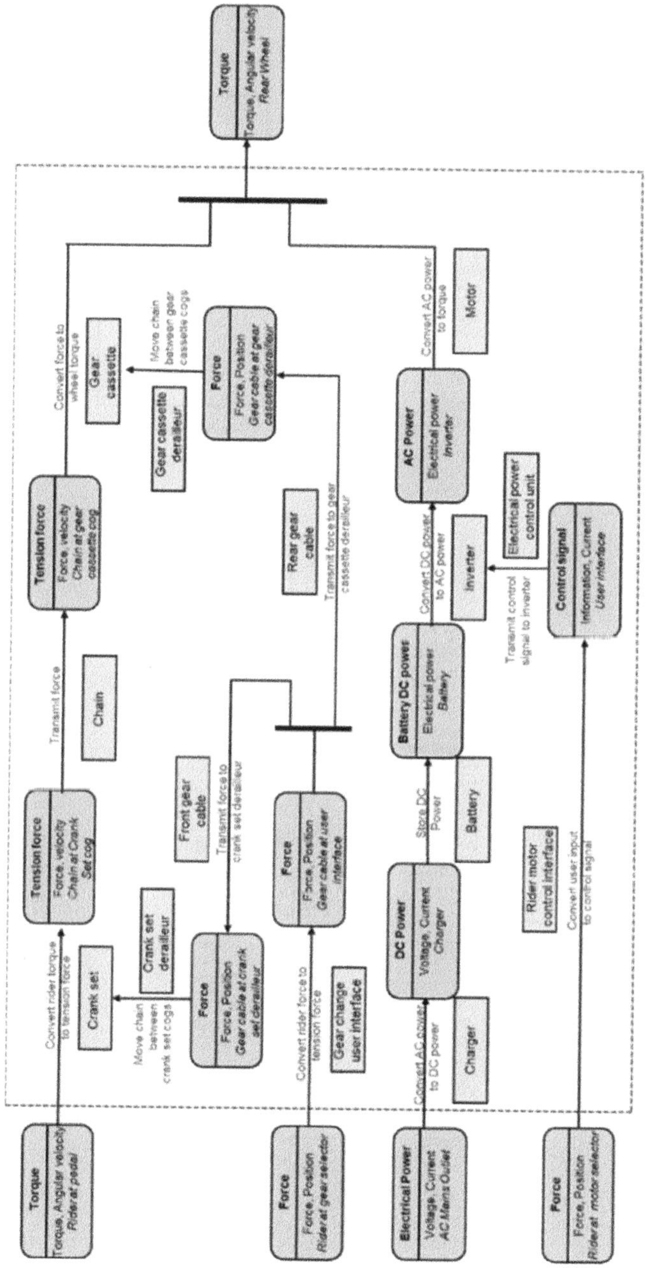

Figure 3.11. *Subsystem-level SSFD for eBike drive system. For a color version of this figure, see www.iste.co.uk/elhami/mechatronics.zip*

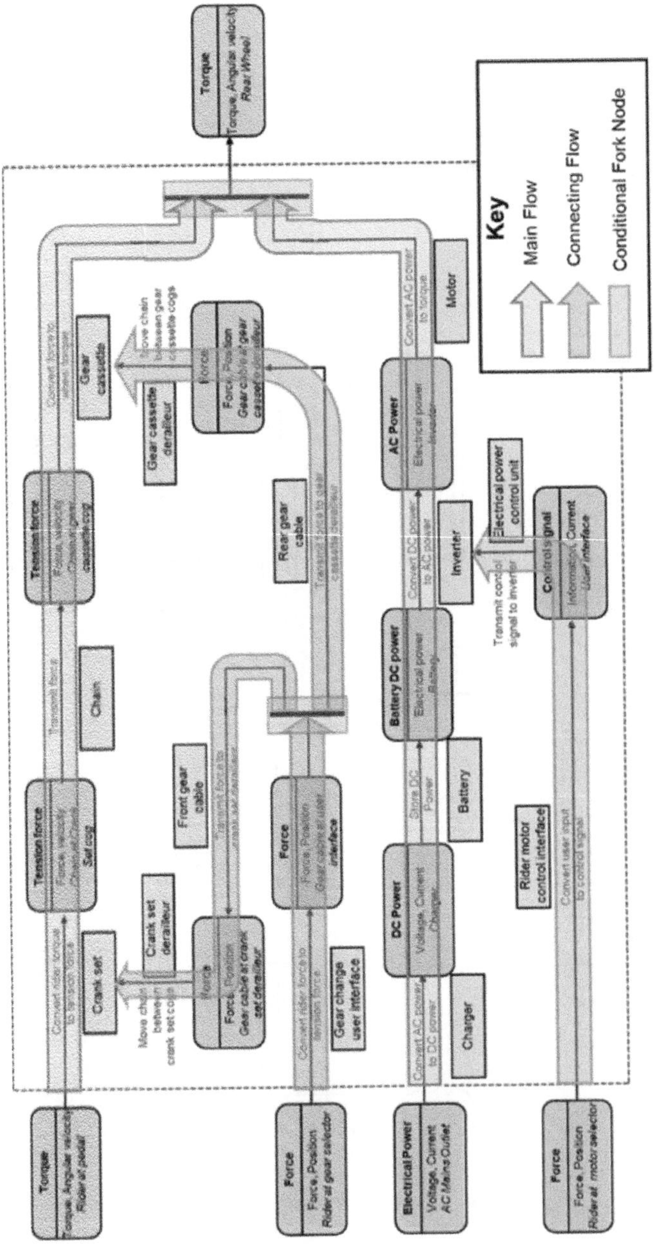

Figure 3.12. *Flow structures identified in subsystem-level SSFD for eBike drive system. For a color version of this figure, see www.iste.co.uk/elhami/mechatronics.zip*

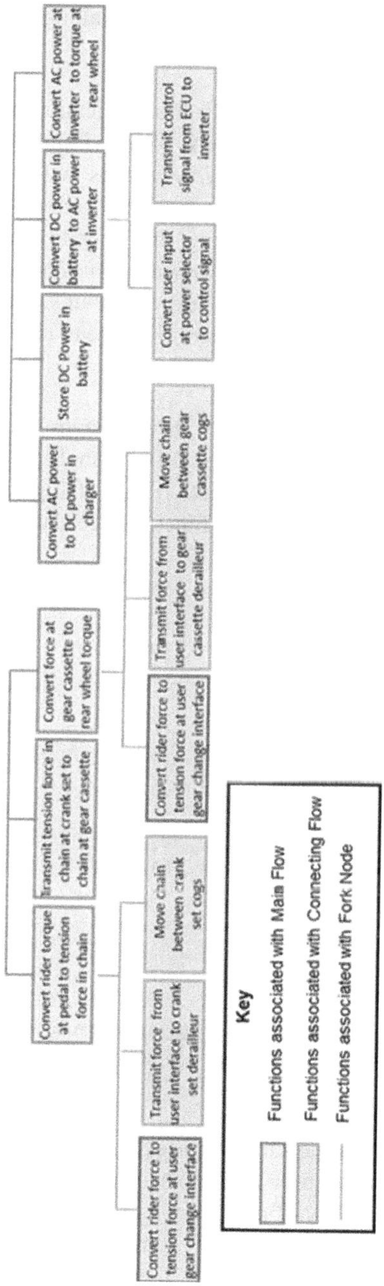

Figure 3.13. *Functions derived from the eBike drive system subsystem SSFD. For a color version of this figure, see www.iste.co.uk/elhami/mechatronics.zip*

The functions defined in the eBike drive system subsystem SSFD and their associated SSFD heuristics are shown in Figure 3.13 along with relationship lines indicating the groupings of the functions.

As Figure 3.13 depicts, in alignment with the application of SSFD heuristics in developing the eBike drive system subsystem SSFD (Figure 3.11), the functional structure is based on two separate *main flow* groups with two *connecting flow* subgroups supporting the operation of the chain set and gear cassette derailleur and a further *connecting flow* subgroup supporting the operation of the inverter. The function *convert rider force to tension force at user gear change interface* appears in both of the human drive system connecting flow subgroups, reflecting the divergent nature of *fork node* associated with the derailleur gear change operation in the SSFD.

Given the relationship between the eBike drive system system- and subsystem-level SSFDs, the subsystem-level functions defined in Figure 3.13 represent the functional decomposition of the system function tree of Figure 3.8 and so the two graphics can be concatenated into the single function tree shown in Figure 3.14.

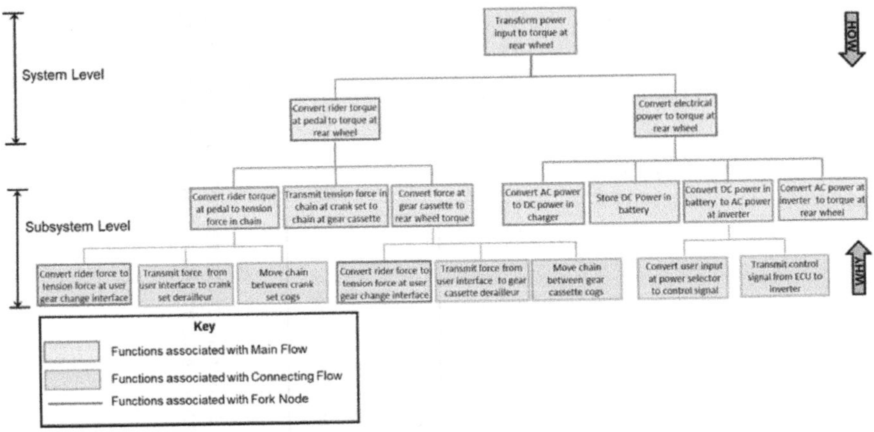

Figure 3.14. *eBike drive system system-/subsystem-level function tree. For a color version of this figure, see www.iste.co.uk/elhami/mechatronics.zip*

At the level of resolution of the subsystem-level eBike drive system analysis, the impact on the system of the rider/user intent to select or deselect the auxiliary motor, that is select the PO or PA modes of operation, is also apparent in the function tree as it was in the corresponding SSFD. The function statements in the function tree include the to/from objects identified on the SSFD which allows the function tree, and any FFT that subsequently develops from it, to stand alone from the SSFD during analysis. Where a system-level SSFD has multiple outputs, it is recommended that a separate function tree is developed for each output to avoid the function tree being overly complex and hence potentially confusing.

The main flow functions at a lower level are connected to the immediately higher-level main flow function through an implied AND gate. For example, in order to achieve the function, *convert rider torque at the pedal to the torque at the wheel* the lower level functions *convert rider torque at the pedal to tension force in the chain, transmit tension force in the chain at the crank set to the chain at the gear cassette* and *convert force at the gear cassette to the rear wheel torque* must all be achieved. There is also an AND relationship for the connecting flow functions, for example, the derailleur gear functions for the crank set. As mentioned earlier, implied AND relationships in a function tree predominate over OR relationships.

The structure of the main flow functions on the function tree can be verified through "how–why" questioning, that is, on a top-down basis any main flow subfunctions immediately below a main flow function on the function tree should answer the question "*How is the high-level function achieved*", equally from a bottom-up perspective the higher-level main flow function should answer the question "*Why is this function required?*" asked of a main flow subfunction. Examples of the question and answer for the eBike drive system function tree are illustrated in Figure 3.15.

The how–why logic may be looser, that is, indirect, for functions associated with connecting and branching flow, for example, where this flow is performance related, that is, enhances rather than being essential to the main flow function that it is associated with. For example, the function of the derailleur gear to move the chain between crank set cogs enhances the main flow function convert rider torque at the pedal to tension force in the chain. Thus, changing gear allows the rider to adopt a cadence more suited to the current road load, but if gear change is inhibited this does not prevent the rider from cycling although it will reduce the overall efficiency of the cycling process. Thus, the question "Why is the chain moved between crank cogs?" is not answered directly by the statement to convert rider torque at the pedal to tension force in the chain but rather by an unstated non-functional performance related requirement such as to make the conversion of rider torque at the pedal to tension force in the chain easier for the rider to accomplish. Any indirect relationship between connecting and branching flow functions and the main flow function they support has implications for an FFT developed from the function tree.

Question		Answer
How is rider torque converted to torque at the rear wheel?	*By*	– Converting rider torque at the pedal to tension force in the chain – Transmitting tension force from the chain at the crank set to the chain at the gear cassette – Converting force at the gear cassette to wheel torque
– Why is rider torque at the pedal converted to tension force in the chain? – Why is the tension force in the chain at the crank set transmitted to tension force in the chain at the gear cassette? – Why is the force at the gear cassette converted to the wheel torque	*To*	Convert rider torque at the pedal to the torque at the rear wheel

Figure 3.15. *How–why question and answer for eBike drive system function tree*

3.3.2. *Subsystem-level function fault tree development*

The development of the FFT is continued at the subsystem level through consideration of the corresponding subsystem-level function tree. Figure 3.16 shows the FFT that has been developed on the basis of the eBike drive system system-/subsystem-level function tree of Figure 3.14.

This time the subsystem-level events immediately beneath the system-level intermediate events are connected to the system-level events by Inclusive OR gates since if any one of the subsystem-level function failure mode events associated with main flow occurs, then the system-level failure mode to which it is linked may occur. That is, in the mirror image of the function tree, the AND gates in the function tree become OR gates in the FFT. The subsystem-level events in Figure 3.16 are also intermediate events since they are not at a level that can be engineered. Being partial function failure events, the Inclusive OR allows for the situation that one, two or more of the subfunction function failure modes may be required for the higher-level function failure to occur. For example, it may be the case that the *low output battery voltage/current for given power input* or the *lower motor output torque for a given AC power input* is not sufficient acting alone to *reduced electrical power converted to torque at the rear wheel* significantly; however, when both lower-level function failure modes occur at the same time, the higher-level function failure occurs.

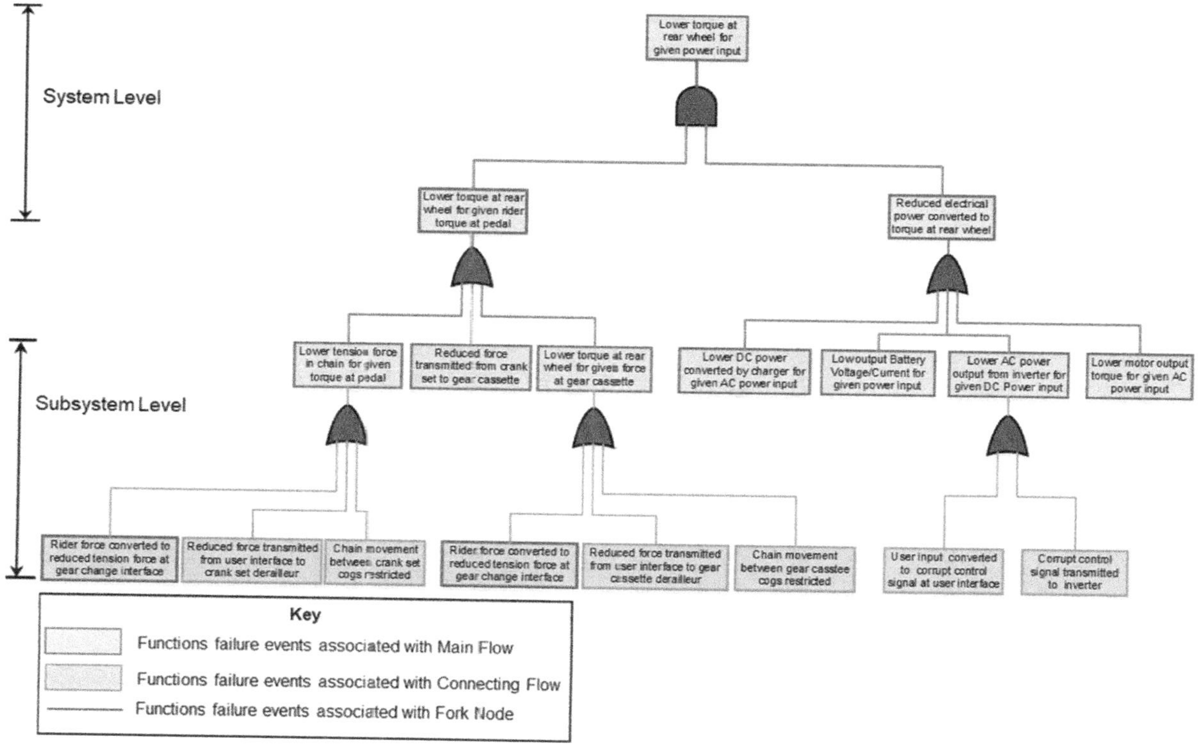

Figure 3.16. *eBike drive system system/subsystem function fault tree. For a color version of this figure, see www.iste.co.uk/elhami/mechatronics.zip*

The function failure events are considered to be independent of each other. In the linear flow of main flow and connecting flow events in the eBike subsystem, the function failure of an upstream event will potentially cause the function failure of downstream events. For example, a *Low output battery voltage/current* may result in *Lower AC power output from the inverter*; however, this is an inevitable result of the physics of the system and not the result of a failure of the inverter. What is important in using an FFT to improve a product design is that the risk of all of the function failure events is eliminated or significantly reduced. That is, the two aforementioned events are both considered independently of each other such that all feasible and probable failure causes are identified so that appropriate design action can be taken to avoid these functional failures occurring in product usage.

3.4. Component-level analysis

3.4.1. *Component-level function decomposition*

At the next system level, in this case at the component level, the eBike drive system functional decomposition has reached a level where the design solutions (components) are typically bought in by the eBike OEM, where the responsibility of suppliers to the eBike OEM is to manufacture these components. It may be advantageous for the eBike OEM to conduct any further functional decomposition by working in conjunction with the part supplier or by the part supplier working alone. At the component level, some of the parts such as those associated with the human drive system, for example, the chain and derailleur, will probably be based on a standard widely used design. Developing an SSFD for such parts may not yield any new knowledge and it may well be more useful in continuing the development of the FFT for these parts to use existing engineering knowledge or, where new knowledge is required, standard engineering techniques such as the mathematical and/or computer-based modeling. Other component-level design solutions such as those associated with the electrically powered aspects of the eBike drive system, for example, the motor, would probably benefit from system state flow analysis. By way of illustration, the eBike motor SSFD is shown in Figure 3.17.

Applying the SSFD heuristics to the development of the SSFD leads to the identification of one *main flow* and two *branching flows* (Figure 3.18).

The functions and SSFD heuristic structures identified in the eBike motor SSFD facilitate component-level function decomposition for the eBike motor, as depicted in the tree structure shown in Figure 3.19.

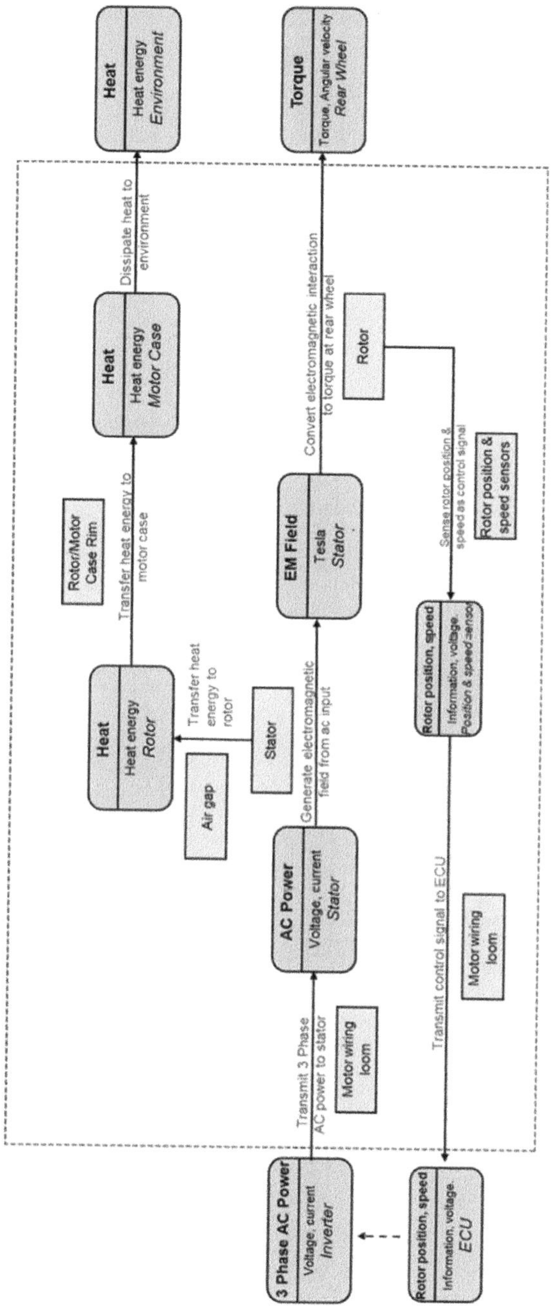

Figure 3.17. *eBike drive system motor SSFD. For a color version of this figure, see www.iste.co.uk/elhami/mechatronics.zip*

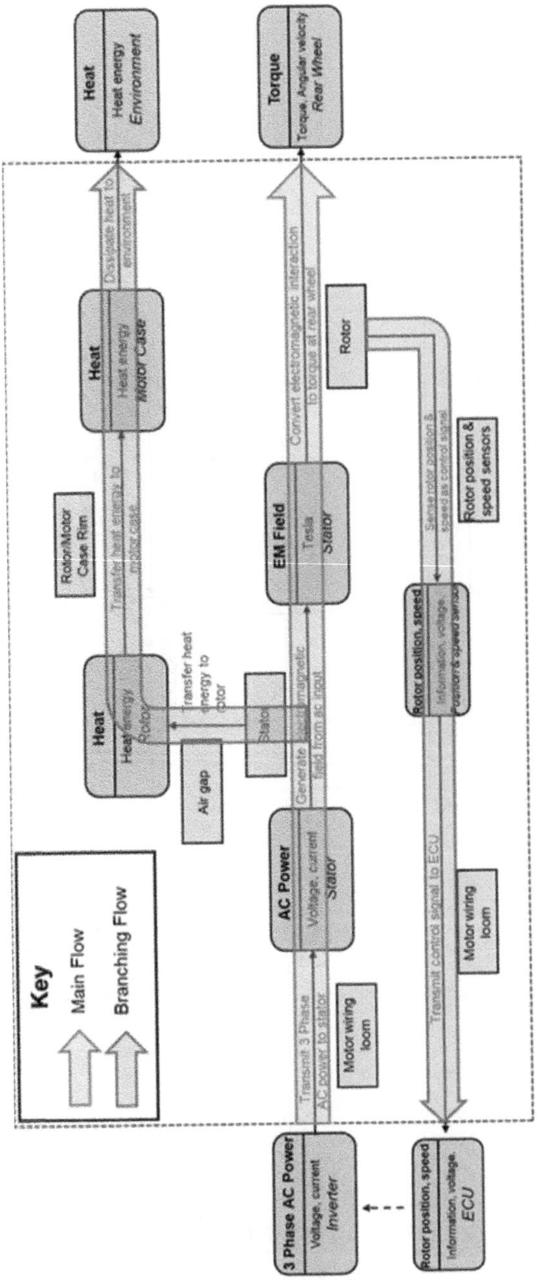

Figure 3.18. *Flow structures identified in motor SSFD for eBike drive system. For a color version of this figure, see www.iste.co.uk/elhami/mechatronics.zip*

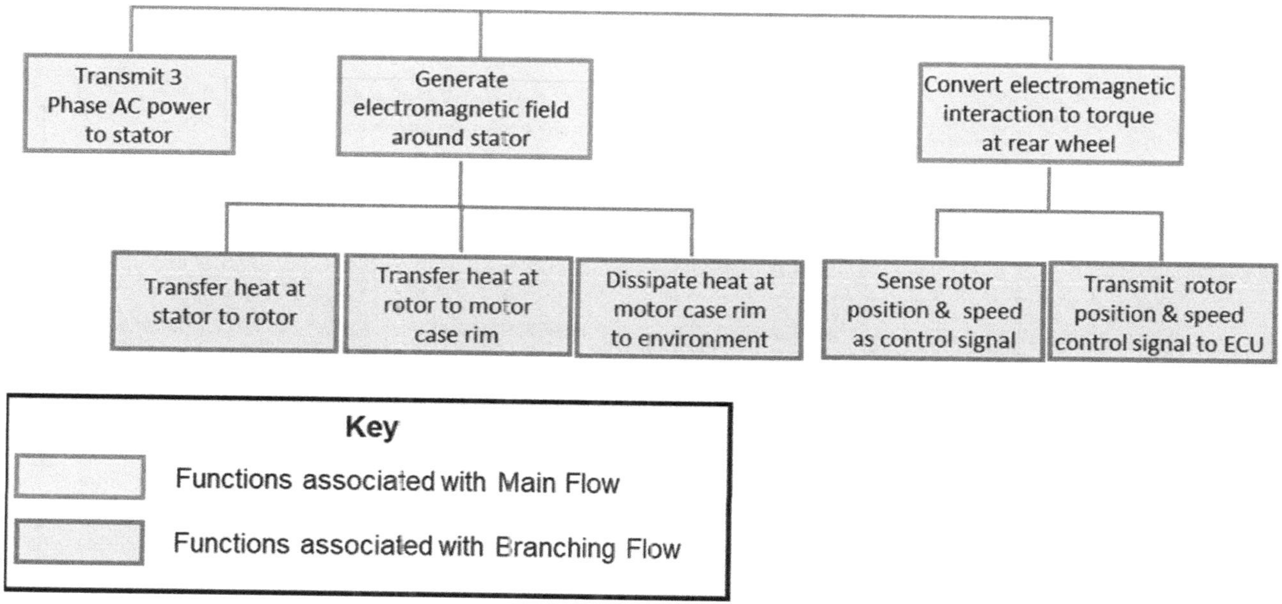

Figure 3.19. *eBike motor function tree. For a color version of this figure, see www.iste.co.uk/elhami/mechatronics.zip*

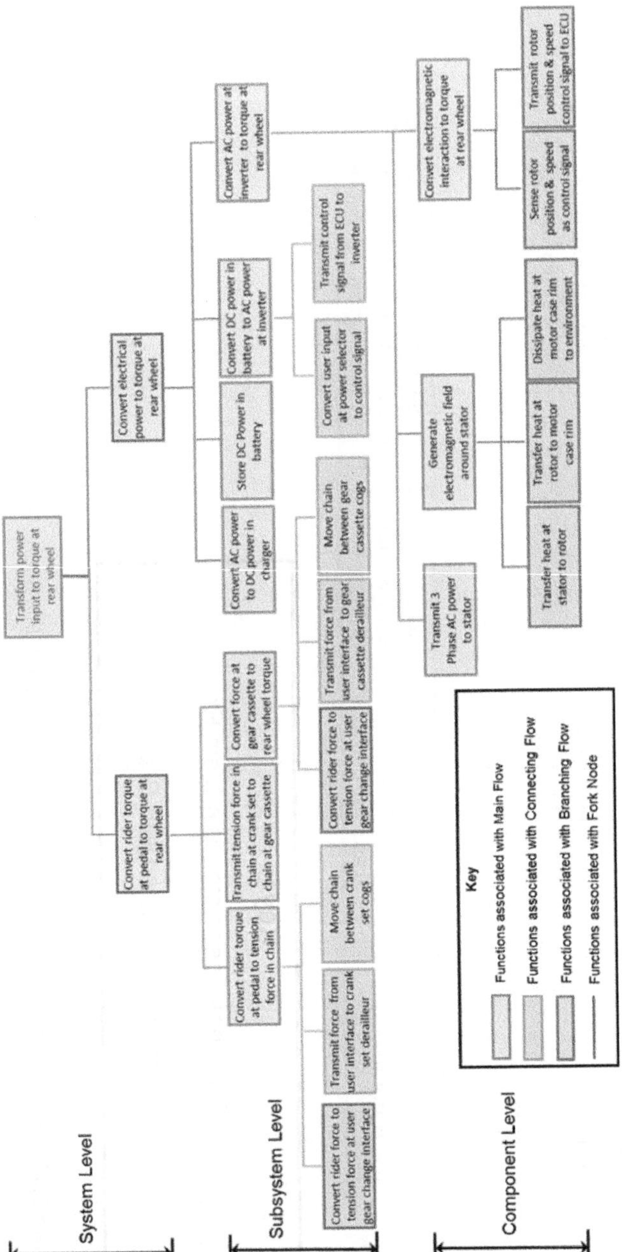

Figure 3.20. *eBike drive system function tree illustrating system-, subsystem- and component-level decomposition. For a color version of this figure, see www.iste.co.uk/elhami/mechatronics.zip*

One of the branching flows is associated with heat flow away from key motor design elements. The functions associated with heat state flow are necessary for good performance of the system, indeed, should these functions not be generated the system will have a significant risk of failure including safety considerations. However, the system can operate without these functions being generated albeit at the risk of less than optimum performance over a much-shortened product lifetime. This means that there is an indirect relationship between the main flow and branching flow functions with the branching functions not being consistent with *how–why* verification of their role in the function tree. As illustrated here, a branching flow often corresponds to “unwanted” energy generated as a result of the main flow.

Contrastingly, the second branching flow associated with sensing the rotor position and speed is essential to the operation of a BLDC motor as this data is a necessary input to the voltage/current software-controlled commutation of the motor by the inverter following input from the ECU. Thus, the functions associated with this branching flow are directly related to the main function that they support.

The eBike motor function tree in Figure 3.19 can be added to the eBike drive system system-/subsystem-level function tree as an example of component-level system decomposition of the tree (Figure 3.20).

The overall structure of the function tree decomposed into three system levels has been influenced by the application of SSFD heuristics in developing the associated SSFDs within each level. In this case, the application of the conditional fork node heuristic at the system level is reflected by the tree being divided into two main branches. The application of the main flow heuristic is reflected by the main groupings of functions at each level. Functional decomposition within a system level includes those functions identified through application of the connecting and branching flow heuristics.

3.4.2. *Component-level function fault tree development*

The system-/subsystem-level FFT can be developed further by taking account of the component-level functional decomposition in the corresponding function tree. The extension of the FFT based on the eBike motor function tree that is shown in Figures 3.19 is illustrated in Figure 3.21.

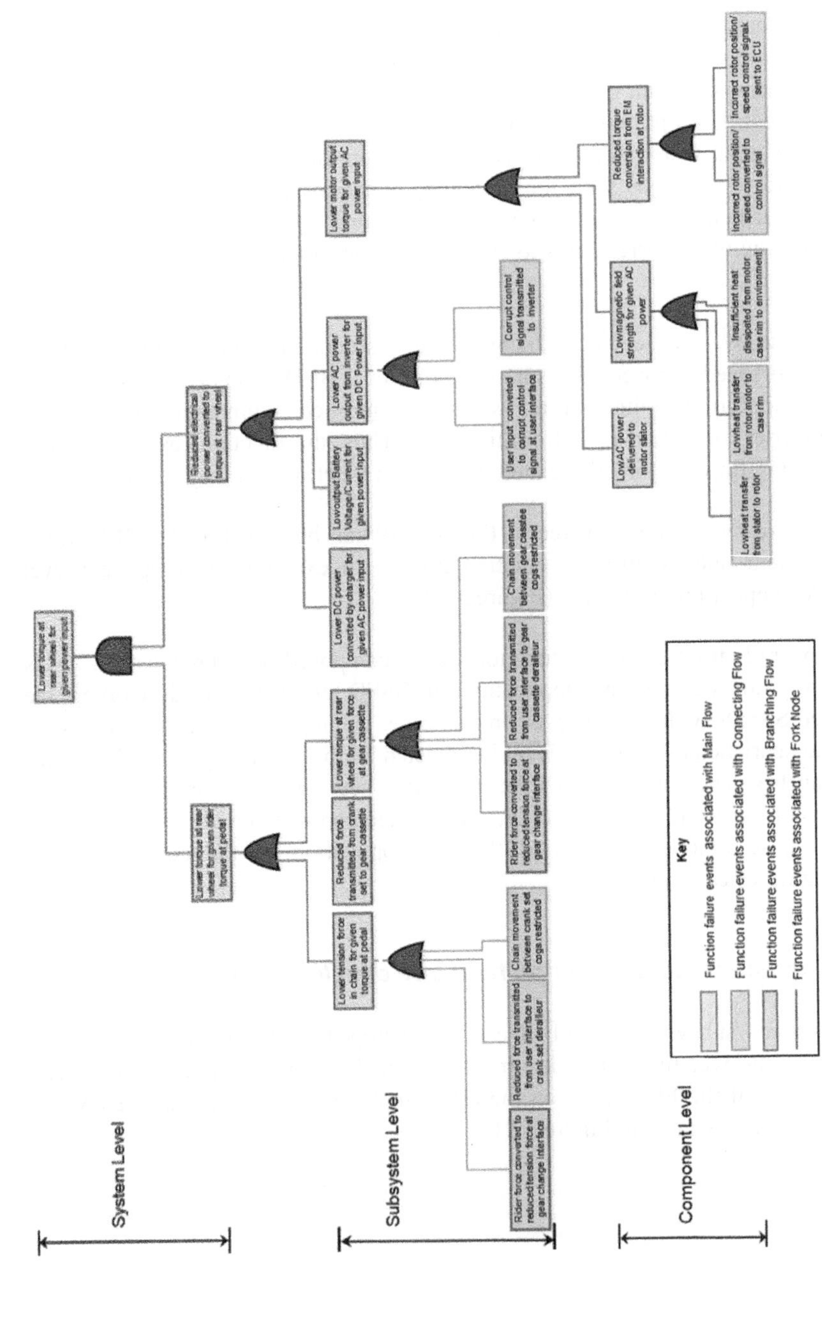

Figure 3.21. *eBike drive system FFT developed to the component-level for the motor. For a color version of this figure, see www.iste.co.uk/elhami/mechatronics.zip*

The FFT of Figure 3.21 now reflects the eBike drive system function tree illustrating the system-, subsystem- and component-level functional decomposition shown in Figure 3.20. A comparison of Figures 3.20 and 3.21 shows that the function tree and its corresponding FFT have the same overall structure. As with the corresponding functions on the function tree the function failure events associated with the application of the main flow, connecting flow and branching flow heuristics during the development of the SSFD are distinguished from each other on the FFT as indicated by the key shown in Figure 3.21. Up to now in developing the FFT the content and structure of the FFT has faithfully mirrored in the failure domain the corresponding function tree set in the success domain.

3.5. Initial analysis of FFT and further decomposition

3.5.1. *Analysis of failure events associated with the connecting and branching flows*

Given that it was not felt useful to decompose the function tree further through SSFD development and the FFT is not yet at a level where design optimization can be fully engineered, the FFT can be decomposed further using engineering knowledge and experience. Attention should be directed initially to those function failure events associated with connecting and branching flows on the FFT to establish whether the link between these failure events and their related higher-level function failure event associated with a main function is *direct* or *indirect*. For example, the linkage of function failure events associated with the derailleur connecting flow to their related main flow function failure event is extracted in Figure 3.22 from the FFT of Figure 3.21.

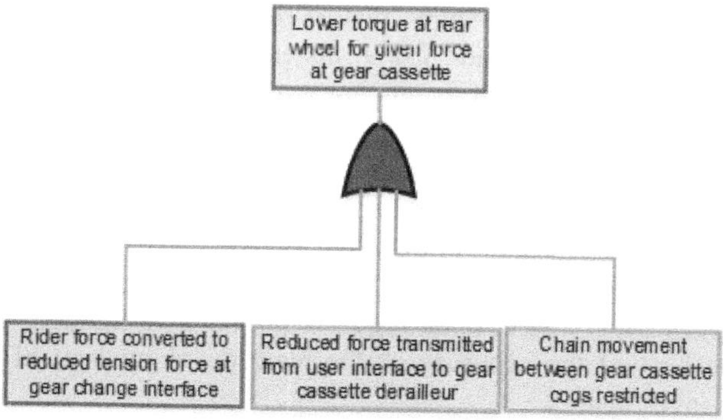

Figure 3.22. *eBike subsystem FFT connecting flow intermediate events. For a color version of this figure, see www.iste.co.uk/elhami/mechatronics.zip*

While all the lower function failure events can cause the higher function failure event, a reduction in the operation of the gear cassette derailleur will not of itself cause lower torque at the rear wheel. That is, the connecting flow events they can only influence the higher function failure event indirectly and so, in this case, the link between the lower connecting flow associated events and the main flow failure event is *indirect*. This means that there must be an additional failure event or multiple failure events that require itemizing, as illustrated in Figure 3.23.

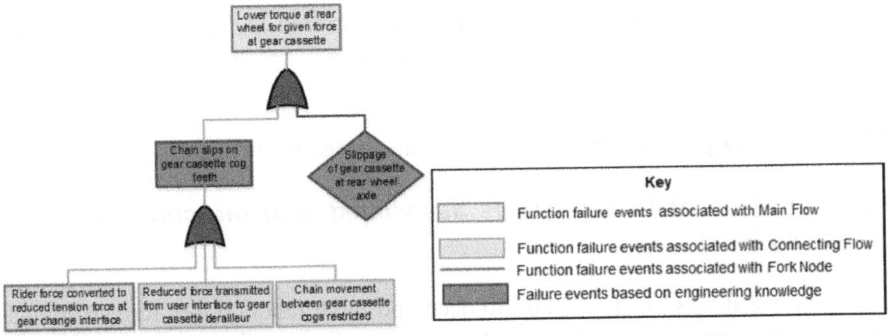

Figure 3.23. *Additional failure events at the chain–gear cassette interface. For a color version of this figure, see www.iste.co.uk/elhami/mechatronics.zip*

The three lowest function failure events associated with the connecting flow are now linked to the highest event through an OR gate and an addition event *chain slips on the gear cassette cog teeth*. This event can be considered the partial function failure mode of the function *chain contacts fully with crank set cogs* with this function being an assumed non-functional (performance-related) requirement. The event is color coded to show that it has not been derived from the corresponding function tree and SSFD instead being based on engineering knowledge and experience. Such events are expressed in conventional fault tree event failure terms rather than being expressed in terms of functional failure. There will be other events that can cause a lower torque to be transferred to gear cassette by the chain (i.e. cause the higher event) such as *slippage of gear cassette at rear wheel axle*. Again, this failure event is associated with a non-functional requirement. This failure event is shown in Figure 3.23 as a diamond, meaning it is an *undeveloped event* and as such requires further investigation. Undeveloped events may be beyond the expertise of the team developing an FFT or require further offline investigation.

Figure 3.24(a), extracted from the eBike FFT of Figure 3.21, shows an example of failure events associated with branching flow with an indirect relationship between lower-level failure events and the higher-level main flow event immediately above them. This indirect relationship was discussed in the context of the corresponding functions on the eBike motor function tree dealing with heat transfer away from the stator and motor. Again, the indirect relationship indicates missing failure events on the FFT as illustrated in Figure 3.24(b).

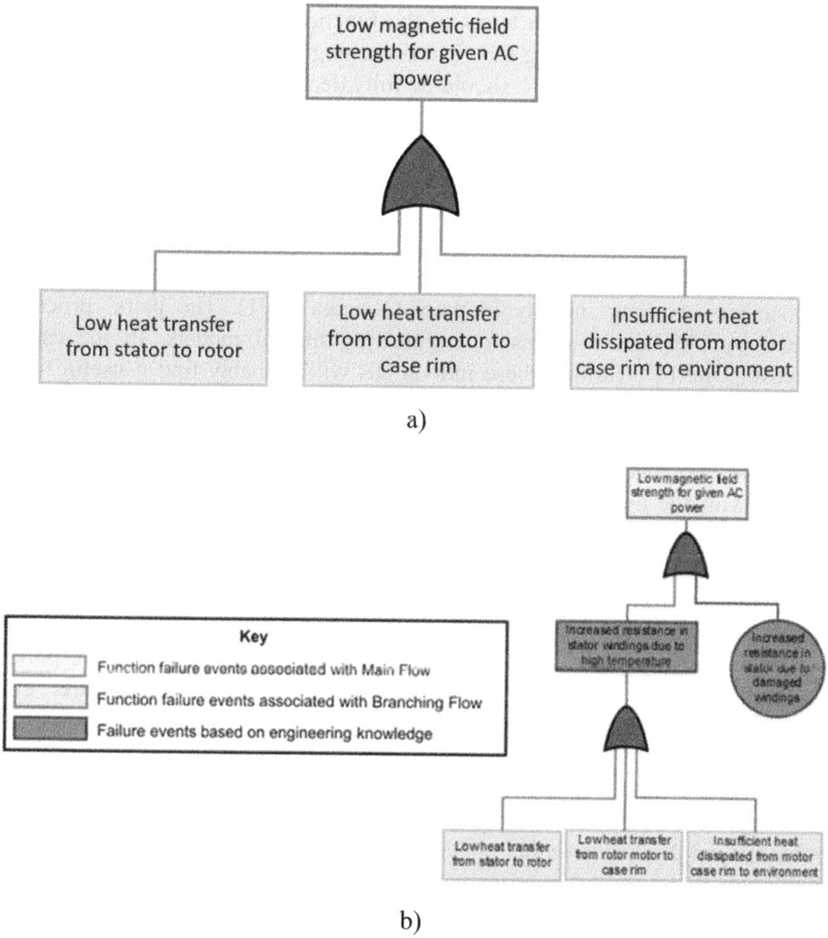

Figure 3.24. *"Missing" failure events in branching flow. For a color version of this figure, see www.iste.co.uk/elhami/mechatronics.zip*

One of the failure events on Figure 3.24(b) is circular, which means that it is a basic event and, as such, is at a level that can be addressed through engineering design.

3.5.2. *Decomposition of the FFT to a level which facilitates design optimization*

Having reviewed the FFT in terms of the function failure events influenced by the application of the connecting flow and branching SSFD heuristics in developing the corresponding SSFD, the FFT should be fully decomposed to the level where the branches of the tree terminate in basic events at which design optimization is facilitated. Any undeveloped events on the FFT will require decomposition to basic events before this can be completed. As mentioned previously, this final level of decomposition of the FFT is best achieved using engineering knowledge and experience. Use should be made of both tacit and explicit engineering knowledge. Tacit knowledge is best captured and disseminated within an engineering team, with team members working initially in pairs (Nonaka 1991). The pairs' process is facilitated by individuals accessing engineering mental models. In the case of decomposing the FFT further, these individuals will probably find it useful to link the lower level function failure events with the design elements depicted on the SSFD which generate the corresponding functions on the associated function tree. For example, the function failure event *low heat transfer from stator to rotor* is associated with the motor stator and motor rotor/case. While generating a mental image of the relative geometry of these design elements, engineers can visualize how this function failure can occur, for example, through physical restrictions to convectional heat flow or by having a rotor with a low heat capacity. In accessing such mental models, it is important to remember that the system under investigation is still in the early stages of design and that the final design may not correspond to the current mental model. Figure 3.25 illustrates function failure event decomposition to a basic event level using tacit and explicit engineering knowledge.

In identifying some basic events, the emphasis may be placed largely on explicit engineering knowledge. The derailleur-based gear change mechanism is a relative complex system that needs further understanding if the aspects of the FFT associated with this are to be decomposed to a basic event level. One way in which such explicit knowledge may be generated at the component level as an alternative to an SSFD was exemplified by Lai and Sung (1998), who developed a mathematical model and a computer simulation of the operation of the rear gear change derailleur mechanism based on the graphical illustration of Figure 3.26. This figure shows four positions (modes) of the chain on the gear cassette in shifting from a higher to a lower gear (smaller to larger diameter cog) over a 170-degree rotation of the gear cassette cogs.

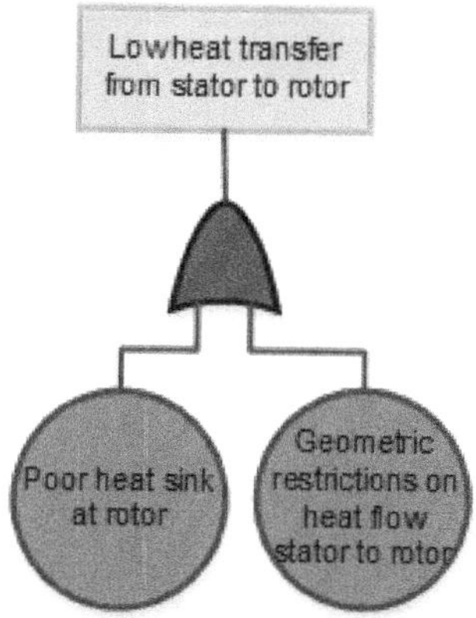

Figure 3.25. *Decomposing of function failure event using engineering knowledge. For a color version of this figure, see www.iste.co.uk/elhami/mechatronics.zip*

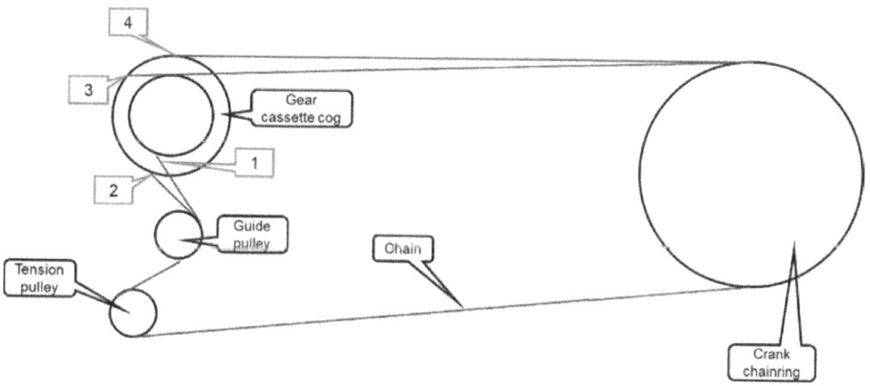

Figure 3.26. *Bicycle chain position on gear cassette cogs (based on Lai et al. 1998). For a color version of this figure, see www.iste.co.uk/elhami/mechatronics.zip*

Figure 3.27 shows basic events additional to those in Figure 3.23 that are typically identified from such work.

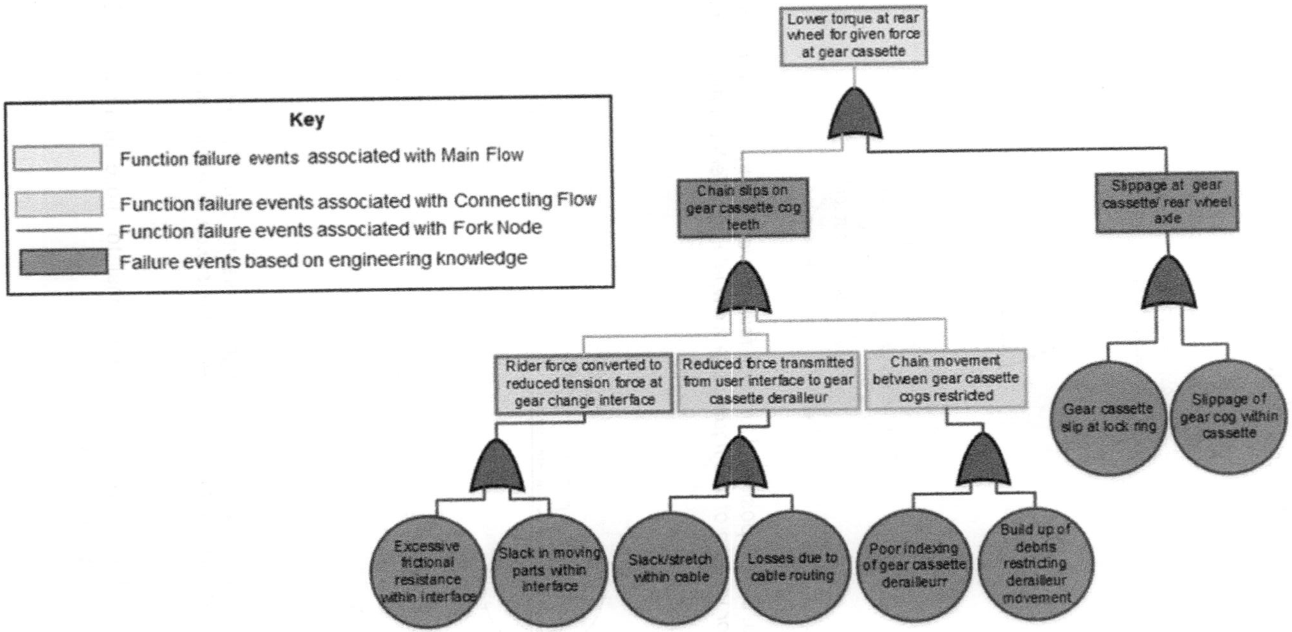

Figure 3.27. *Identification of further basic events using engineering knowledge. For a color version of this figure, see www.iste.co.uk/elhami/mechatronics.zip*

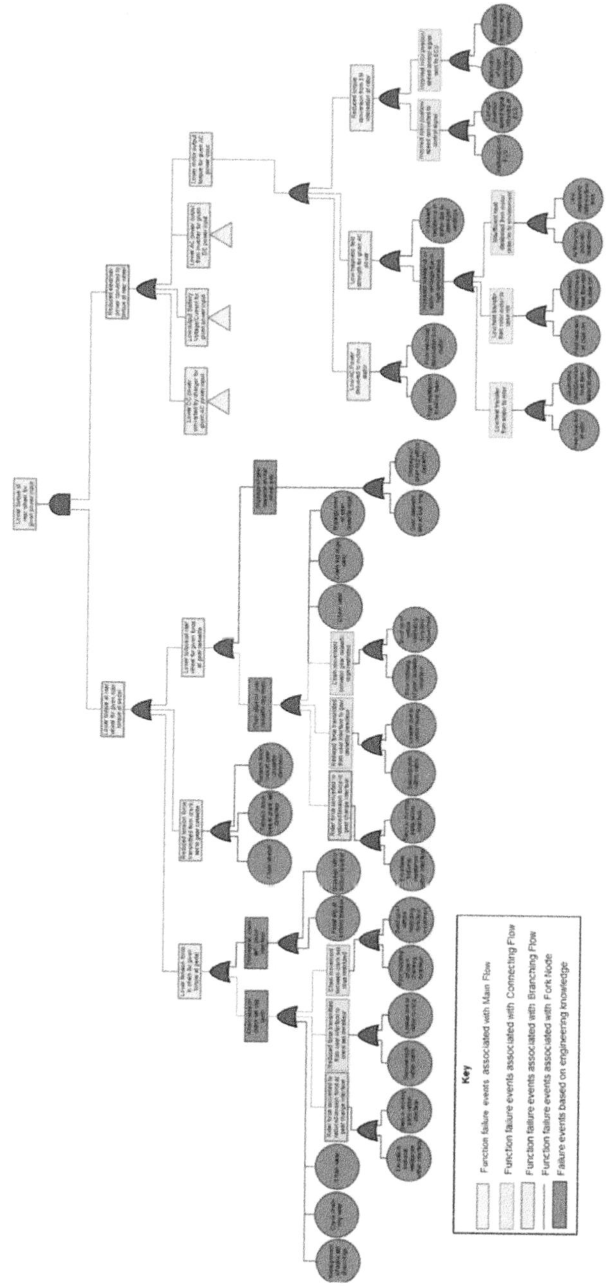

Figure 3.28. *Part of eBike drive system FFT developed to the basic event level. For a color version of this figure, see www.iste.co.uk/elhami/mechatronics.zip*

In Figure 3.28, additional failure events identified from engineering knowledge have been added to the FFT to decompose the eBike drive system FFT to a level that facilitates design optimization; not all potential basic events are shown on the FFT to restrict the size of the tree that is illustrated.

3.6. eBike drive system function fault tree analysis

Fault tree analysis (FTA) is widely used in a quantitative manner to predict system reliability by computing the probability of system failure given the probability of failure of its component parts. However, this can prove difficult when the methodology is used during the design stage of a system since failure rate data is probably not available (Campean and Henshall 2008). This is especially the case for a new design or an existing design which incorporates new features, particularly where these new features are based on a different technology to the existing design. In addition, with a new design, an FFT developed early in the design process may well include undeveloped events or transfer gates relating to parts of the FFT or separate FFTs that are still being developed. This can be because the subsystems or components associated with these events and trees may themselves still have to be designed or, if they are to be bought in as off-the-shelf parts, they have yet to be selected since their requirements specification is still being developed.

The aim of any FTA conducted as an integral part of product design is (i) to identify failure events that will potentially have a detrimental impact on system dependability, that is, on the safety, reliability, availability and maintainability of the system, and (ii) to develop appropriate design based countermeasures as required to ensure the system dependability is not compromised in customer usage. Complementary methods for the analysis of an FFT are now considered along with illustrations of countermeasure design strategies that may be taken, namely: design concept choice, design optimization, functional and hardware redundancy, compensation and upgrading of materials. Thus, the methods of analysis can be categorized as:

- macro-level analysis;
- analysis based on SSFD flow heuristics;
- analysis based on cut sets.

3.6.1. *Macro-level function fault tree analysis*

In the case of an FFT developed as the mirror image of a function tree derived from an SSFD in which SSFD heuristics were applied, the initial analysis should focus on any patterns (substructures) present in the FFT. The influence of the SSFD heuristics on the macro-structure of the eBike drive system FFT is illustrated in Figure 3.28 by the color coding of the failure events and their relationship lines.

The first striking feature of the tree is that it is divided into left-hand and right-hand sections corresponding to the human and electric drives respectively making the influence conditional fork node heuristic at the system level readily apparent. The structure of both sections of the tree is heavily influenced by the main flow heuristic. The connecting flow heuristic also has an important influence on the overall number of failure events in the human drive section of the tree and the branching flow having a similar influence on the electric drive section. Due to its top-down nature, the tree retains the structure determined by the influence of the SSFD heuristics even though approximately two-thirds of the lower-level failure events present on the FFT are based on engineering knowledge and experience. This means that the overall structure of the FFT reflects the operational flows of material, energy and information in the system being analyzed enabling failure events on the tree to be interpreted in the context of system operation.

There are three triangular transfers in symbols on the eBike electric drive system section at the subsystem level of analysis in Figure 3.28, indicating that these aspects of the tree are continued elsewhere on a separate page or tree. This type of separation of the FFT is considered in more detail later. The human drive part of the tree is fully decomposed to a basic event level although the tree does not show all failure events to maintain clarity of the graphic. Identification of function failure events associated with SSFD heuristic flows from a macro-perspective provides a powerful strategy for countermeasure development to address the effect of single or groups of basic events and is considered next.

3.6.2. *Function fault tree analysis based on SSFD heuristics*

The analysis of the FFT based on SSFD heuristics can be conducted at each of the system levels in sequence as the FFT is developed as an integral part of the product design process.

System-level flow analysis

The system-level SSFD with SSFD heuristic flows and the corresponding partial function failure mode FFT are shown alongside each other in Figure 3.29.

The FFT reflects the combination of the human and electric drive main flows at the fork node in the SSFD by the decomposition of the FFT top event into two system-level intermediate events. Given that the energy flows combine, it is possible that one or both flows could have a detrimental effect on the other. An example of such an unwanted interaction is one that could result with a mid-drive motor (one that is located alongside the pedals and the crank set) where the human and electric drives share the same transmission to the rear wheel putting additional stress on the human drive system components (chain and gearing). This will result, for example, in failure events on an FFT for a drive system based on this design concept relating to increased chain wear. In other words, the choice of design concept at the system level is one countermeasure strategy that can be employed to address potential failure events which would otherwise become apparent at the component level and have important implications for system dependability. As a result of system-level analysis, a BLDC rear hub direct drive motor was chosen. Brushless motors are inherently more reliable and efficient than brushed motors.

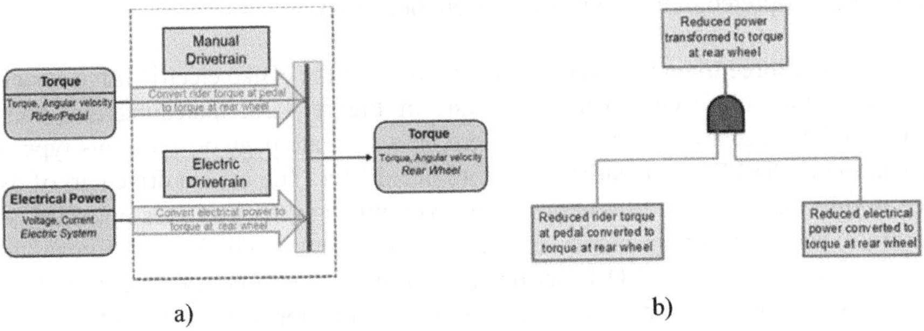

Figure 3.29. *eBike drive system system-level analysis. (a) eBike drive system system SSFD and (b) eBike drive system system FFT. For a color version of this figure, see www.iste.co.uk/elhami/mechatronics.zip*

As mentioned earlier, the term *reduced* requires quantification, that is, what is the threshold level of reduction of human and/or electrical power beyond which the performance of the eBike is no longer acceptable. This imposes a condition on the FFT which can be documented graphically by the inclusion of a special type of AND gate termed an INHIBIT gate in which the inputs to the gate include the condition which is imposed on the gate. Figure 3.30 shows the condition imposed on the eBike system-level FFT of Figure 3.29(b) stated in the form of a transfer function.

The transfer function will model the power output of the system to the eBike rear wheel as a function of the human and electrical power inputs and so quantify an acceptable level of output in terms of the input levels. The analytical transfer function developed early in the product design process can subsequently be realized within the eBike ECU coding along with signal flow from sensors measuring input and output power. In order not to increase the complexity of the case study, this aspect is not considered in the analysis that follows.

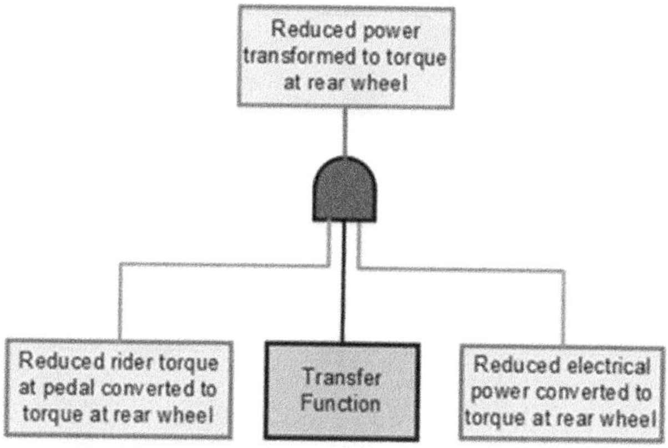

Figure 3.30. *eBike system-level FFT with inhibit gate. For a color version of this figure, see www.iste.co.uk/elhami/mechatronics.zip*

Subsystem-level flow analysis

The development of an FFT on the basis of its corresponding fault tree which has been derived from an SSFD to which SSFD flow heuristics have been applied ensures that the structure of the FFT is consistent with the physics of the operation of the system. The structural heritage of the FFT can be exploited in an analysis of the FFT as is illustrated next by two examples of subsystem-level FFT analysis that uses design optimization to develop countermeasures to failure.

Figure 3.31 highlights with broken line boxes the subsystem-level failure events directly associated with SSFD flow heuristics in the human drive system section of the eBike drive system FFT that are focused on in the example that follows.

Figure 3.32(a) shows the connecting flow for the crank set derailleur extracted from the corresponding subsystem-level SSFD (Figure 3.12) while Figure 3.32(b) shows the intermediate events and linked basic events that correspond to this flow extracted from Figure 3.31.

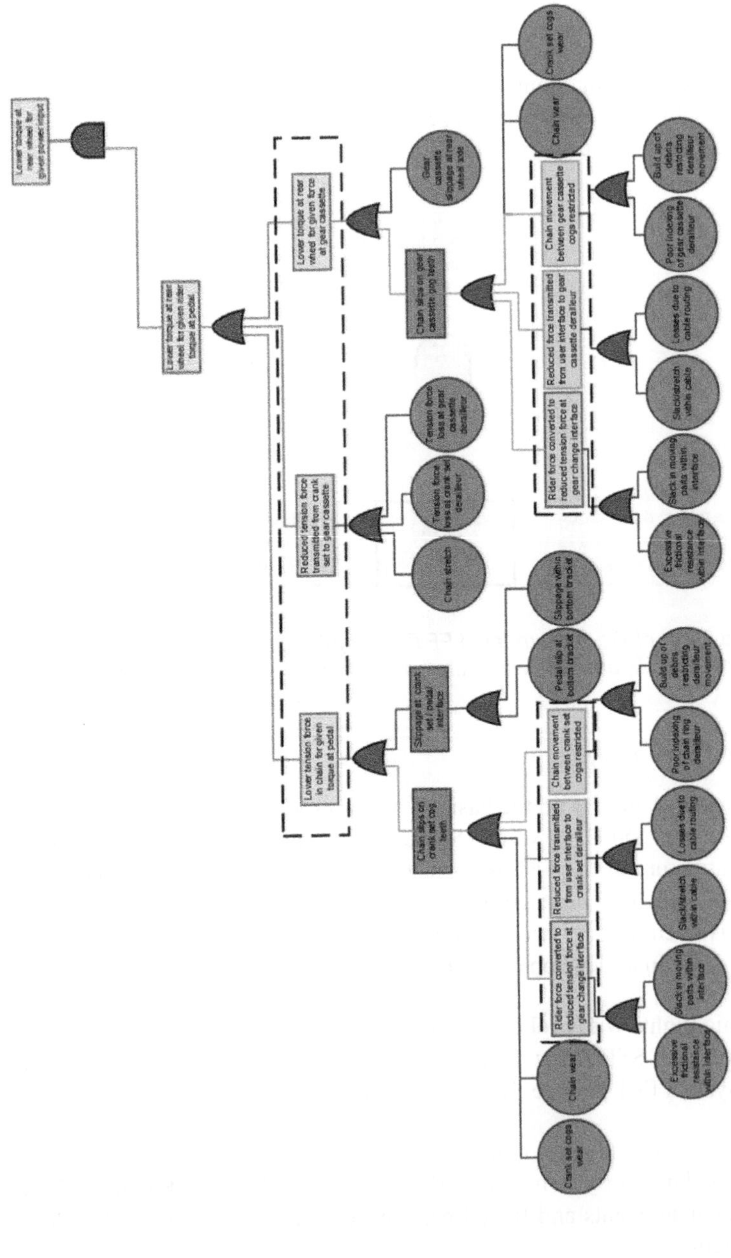

Figure 3.31. *Failure events associated with SSFD heuristic flows. For a color version of this figure, see www.iste.co.uk/elhami/mechatronics.zip*

By definition, the basic events linked to an intermediate event through an OR gate cause the partial function failure mode stated in the intermediate event. For example, *slack/stretch within cable* and *losses due to cable routing* both result in *reduced force transmitted from user interface to crank set derailleur*. Assuming a model of the energy flow (force and resulting movement) associated with the connecting flow exists, or can be readily developed, the effect of the basic events can be included in the model as a reduction of the force and movement. The combination of basic events should indicate the magnitude of the effect to be included in the model; where this is not known from previous analysis, it will need to be estimated with a view to subsequent confirmation. Assuming that the model includes design parameters that can be manipulated, it can be used to optimize the design to be robust to the effect of the basic events, that is, to the basic events themselves. A limit will need to be set on the magnitude of the basic event effects associated with regular maintenance of the human drive system, for example, cable stretch. A review of the human drive system section of the eBike drive system FFT of Figure 3.28 shows that the crank set and gear cassette derailleur connected flow intermediate and linked basic events are essentially the same and so the analysis for the crank set derailleur can largely be carried across to the gear cassette derailleur.

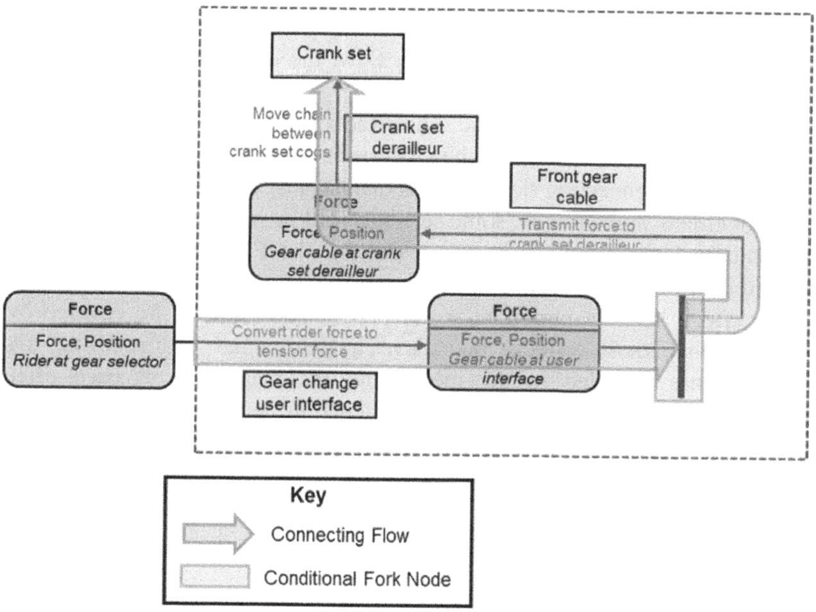

Figure 3.32. *a) Extract from eBike drive system subsystem SSFD. For a color version of this figure, see www.iste.co.uk/elhami/mechatronics.zip*

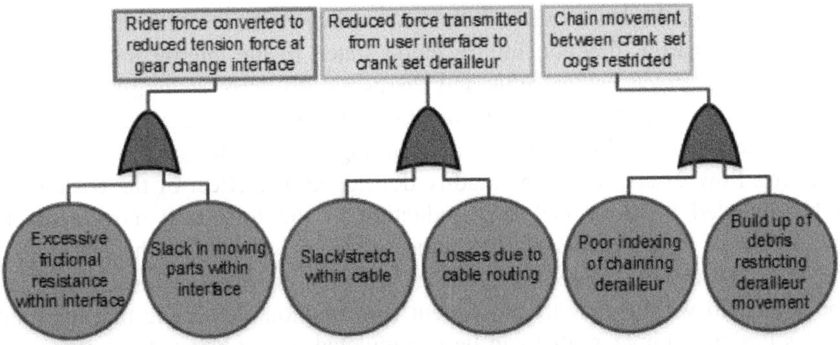

Figure 3.32. *b) Extract of failure events from eBike drive system FFT. For a color version of this figure, see www.iste.co.uk/elhami/mechatronics.zip*

Continuing the subsystem-level flow analysis of the eBike human drive system FFT, the subsystem main flow of the eBike human drive system is considered next. The corresponding extracts from the eBike drive system subsystem SSFD and FFT are shown in Figure 3.33.

As with the analysis of the connecting flow, the effect of basic events can be included in any model of the main flow of energy/force so that the model can be used to optimize system performance. The effect of the basic events associated with the connecting flows is included in the analysis as a single entity for each derailleur flow based on the connecting flow analysis that has just been described. In including basic event effects in the model, some selection will be necessary, particularly given the potentially large number of such events on the FFT. This selection can be conducted based on the effect of basic events as multiple events are likely to have the same effect. What is important is not that all basic events are covered by the model, but that those effects that have a detrimental influence on system performance are included in the model and the magnitude of each effect is representative of what the system will experience in operation. The system design needs to ensure that the performance of the system is robust to the detrimental effects.

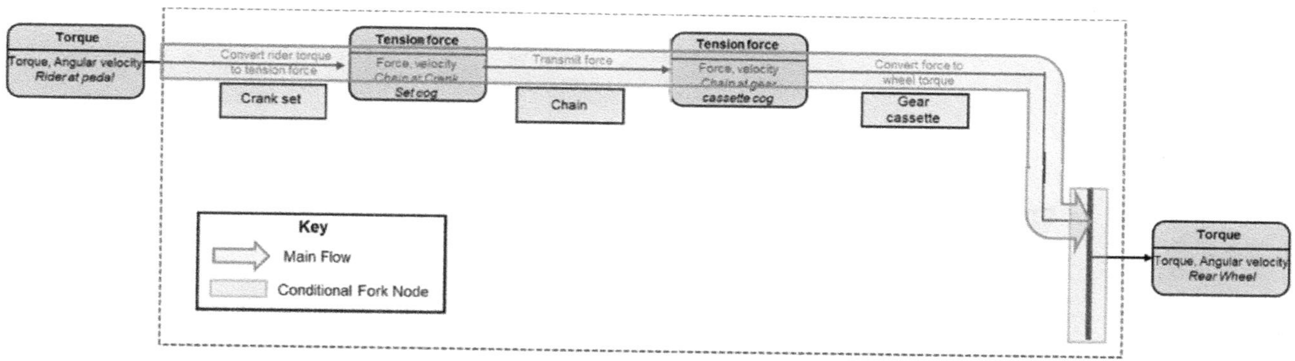

Figure 3.33. *a) Extract from eBike drive system subsystem SSFD. For a color version of this figure, see www.iste.co.uk/elhami/mechatronics.zip*

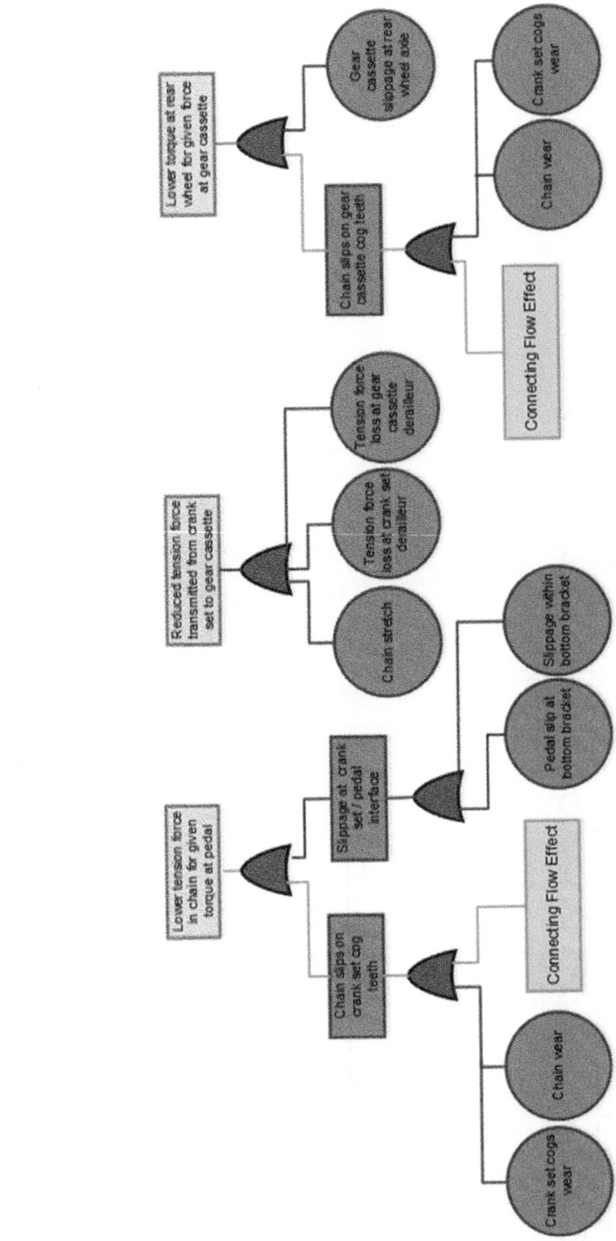

Figure 3.33. *b) Extract of failure events from eBike drive system FFT. For a color version of this figure, see www.iste.co.uk/elhami/mechatronics.zip*

Component-level flow analysis

There will be some basic events acting individually or in combination with others that have a significant effect on system dependability such that design optimization is not an appropriate strategy. Again, intermediate events corresponding to SSFD heuristic flow offer a good insight into how countermeasures may be developed. Figure 3.34 shows the electric drive system section of the eBike drive system FFT with the failure events associated with an SSFD heuristic branching flow, highlighted with a dotted line box, that are considered next at this level of analysis.

Figure 3.35 shows extracts from the eBike drive system subsystem SSFD and FFT corresponding to the branching flow highlighted in Figure 3.34 associated with the detection of the motor rotor speed and position.

Sensing of the rotor position and speed is vital to the operation of the BLDC motor since software commutation of the motor is dependent on this information. Any of the basic events in Figure 3.35(b) acting alone can potentially stop the motor operating and so may be considered safety critical. One countermeasure that can be employed is to use functional redundancy in which the functions associated with the branching flow are achieved using a different technology alongside that using the rotor position and speed sensors. For example, commutation without the need for sensors can be achieved by detecting the back-EMF induced in the rotor as it moves into the idle phase, that is, moves through the position where commutation is required (Zeng and Li 2011). This sensorless control requires extra circuitry and a different control algorithm.

Hardware redundancy can also be used as a countermeasure by doubling up on the number of sensors with duplicate information paths to the ECU. The disadvantage of using redundancy is increased cost, weight and packaging and so this type of countermeasure is generally restricted to safety critical situations. Functional redundancy generally has increased reliability over hardware redundancy since the conditions that cause one set of hardware to fail can also cause the backup hardware to fail. Being functionally based, the FFT is particularly powerful in detecting opportunities for design countermeasures to failure based on functional redundancy.

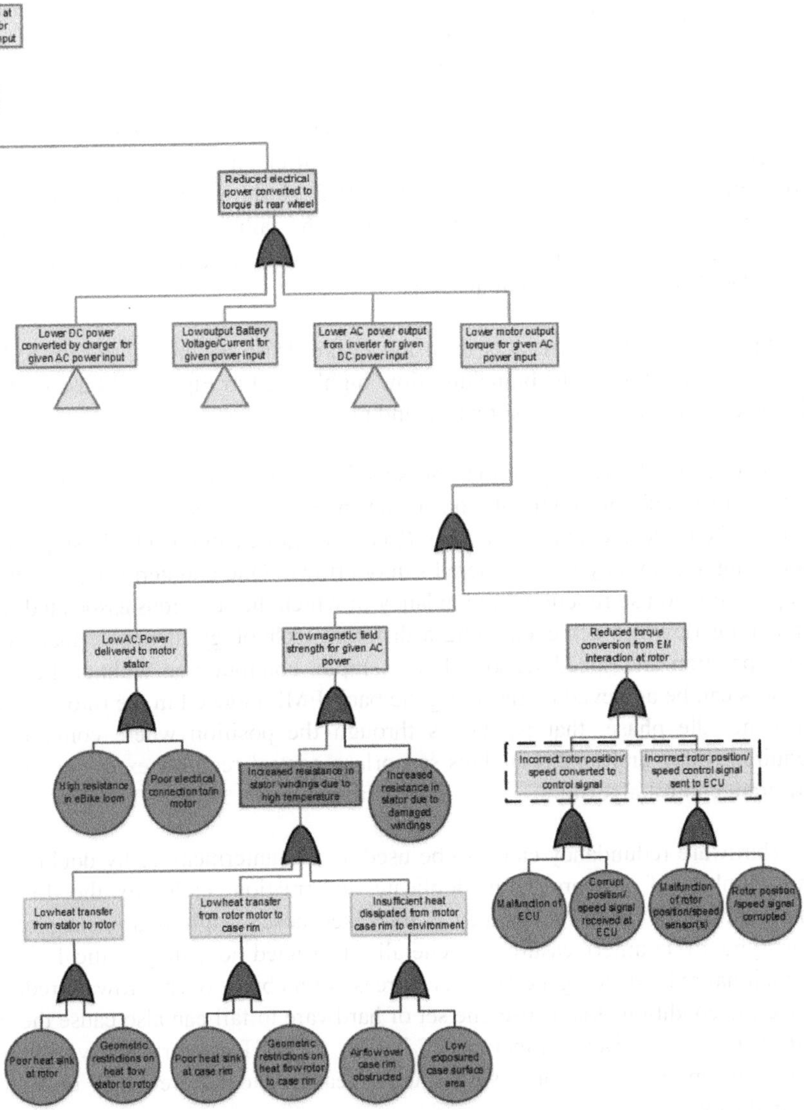

Figure 3.34. *Electric drive system – highlighted failure events associated with a flow heuristic. For a color version of this figure, see www.iste.co.uk/elhami/mechatronics.zip*

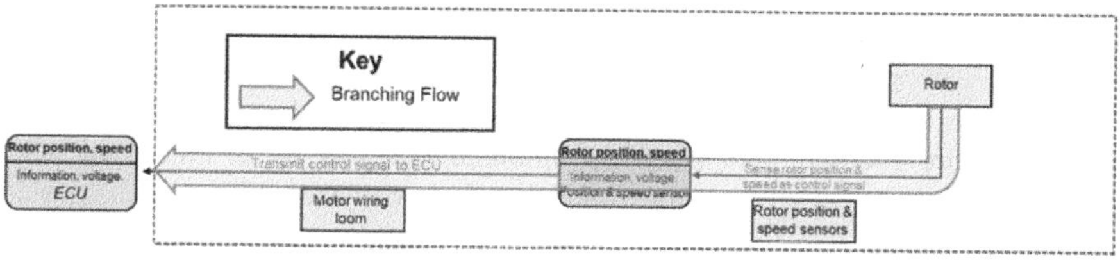

Figure 3.35. *a) Extract from eBike drive system subsystem SSFD. For a color version of this figure, see www.iste.co.uk/elhami/mechatronics.zip*

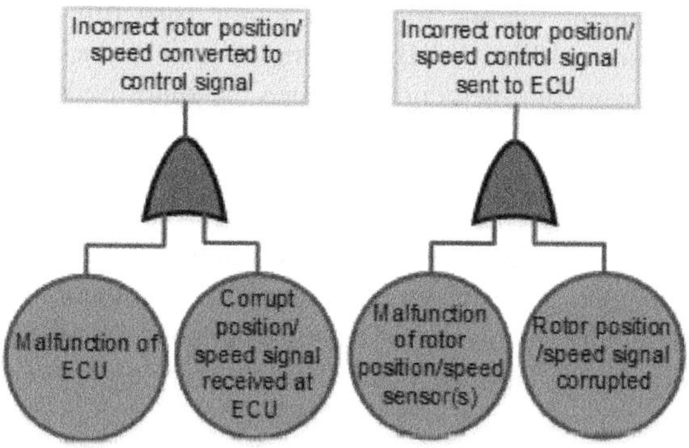

Figure 3.35. *b) Extract of failure events from eBike drive system FFT. For a color version of this figure, see www.iste.co.uk/elhami/mechatronics.zip*

3.6.3. *Function fault tree analysis based on cut sets*

The FFT can be analyzed in the conventional manner for a fault tree by identifying cut sets and minimal cut sets; a cut set being a combination of basic events that can cause, that is, have a direct link to, the top event while a minimal cut set is the smallest combination of basic events that can result in the top event (Vesely 2002). The FFT shown in Figure 3.28 has a top event that is a partial function failure mode with all the intermediate and basic events also representing partial function failure modes. As considered previously, the Inclusive OR linking the causal events allows for the situation that one, two or more of the lower failure events may be required for the higher-level function failure to occur. Thus, it is equally important to identify combinations of partial function failure events as single partial function failure events acting alone that may cause the top event and hence system failure. It is also the case that even in a moderately complex FFT, there are a high number of possible combinations of partial function failure events that potentially can cause the top event. In this situation, the authors recommend analyzing the FFT both from the more holistic perspective based on cut sets and from the more focused perspective of minimal cut sets as is illustrated next. Figure 3.36 shows a cut set from the eBike electric drive system FFT.

Being associated with a branching flow in the corresponding eBike electric drive system SSFD, the basic events in the cut set of Figure 3.36 reflect a holistic perspective of heat flow in a BLDC motor. The main source of heat is due to the electrical resistance of the stator windings with the primary heat flow being from the stator to the environment via the rotor and motor case rim (Chai et al. 2016). Increasing temperature within the BLDC motor will reduce its efficiency by reducing the electromagnetic interaction between the stator both through a decrease in current flow in the stator windings and a decrease in magnetic field density in the permanent magnet of the rotor. Any reduction in motor efficiency will reduce the overall range that a fully charged battery will give. The basic events point to the need to optimize the flow of heat away from the stator windings to the case rim and hence to the environment. A design strategy based on the modeling of this flow can be effective in minimizing the risk to the customer of poor perceived performance of the system in terms of range between battery charging by enabling a better understanding to be gained of the effect on heat flow of design parameters such as the geometry and composition of the stator and the rotor as well as the geometry of the air gap between the stator and rotor.

Continuing the FFT analysis based on cut sets, Figure 3.37 shows an example of a minimal cut set from the human drive system section of the FFT for a top event corresponding to total function failure.

The minimal cut set illustrates that the basic event *slack/stretch within cable* acting alone can cause the top event *no torque at rear wheel*. Note that this minimal cut set includes total function failure events. The other basic events are grayed out, not being a part of this minimal cut set, but each could potentially form its own minimal cut set. The counter measure strategy which is typically applied to this basic event is compensation in which there is an adjuster within the cable length that can compensate for a certain amount of cable stretch within the limits of regular system maintenance. Should the design specification for the cable not be such as to keep the cable stretch down to a limit acceptable within a routine maintenance schedule, then the additional countermeasure strategy of upgrading the material could be implemented.

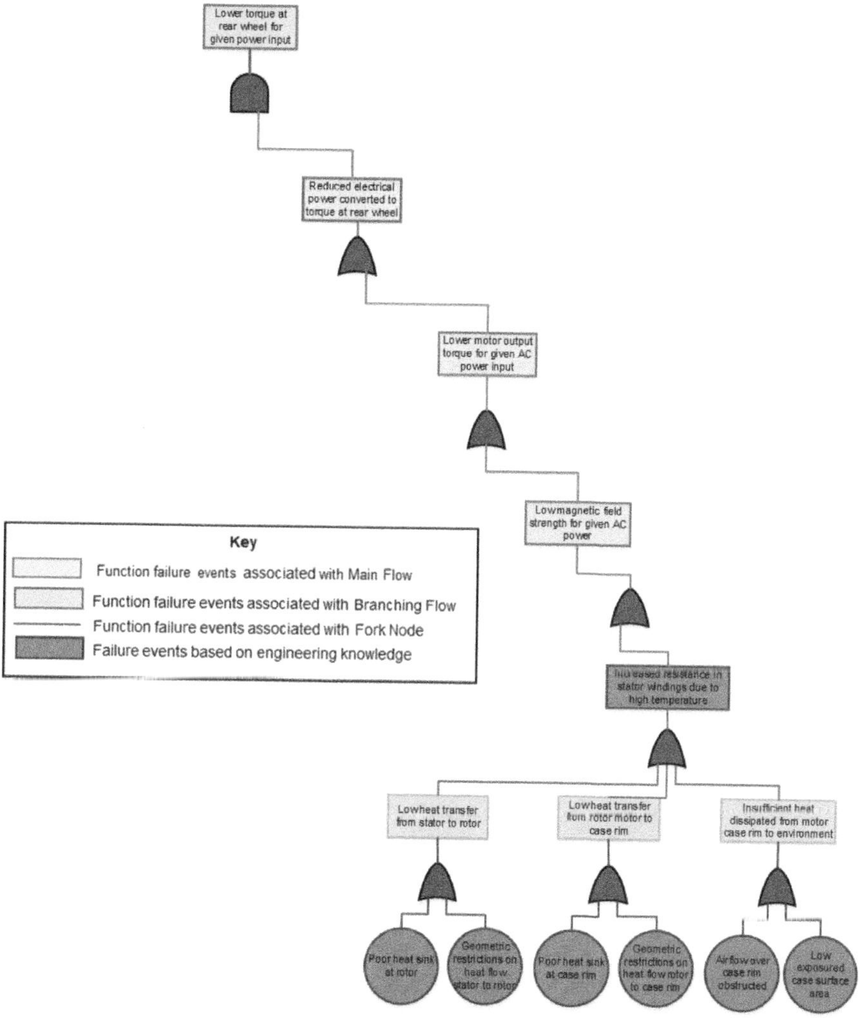

Figure 3.36. *Electric motor cut set. For a color version of this figure, see www.iste.co.uk/elhami/mechatronics.zip*

This example of a total function failure top event highlights that, although the focus of the FFT development and analysis in this chapter has been on partial function failure modes, the other types of functional failure modes of no function, intermittent function and unintended function should also be considered during the product design process when a failure mode type is feasible and has a reasonable probability of occurrence.

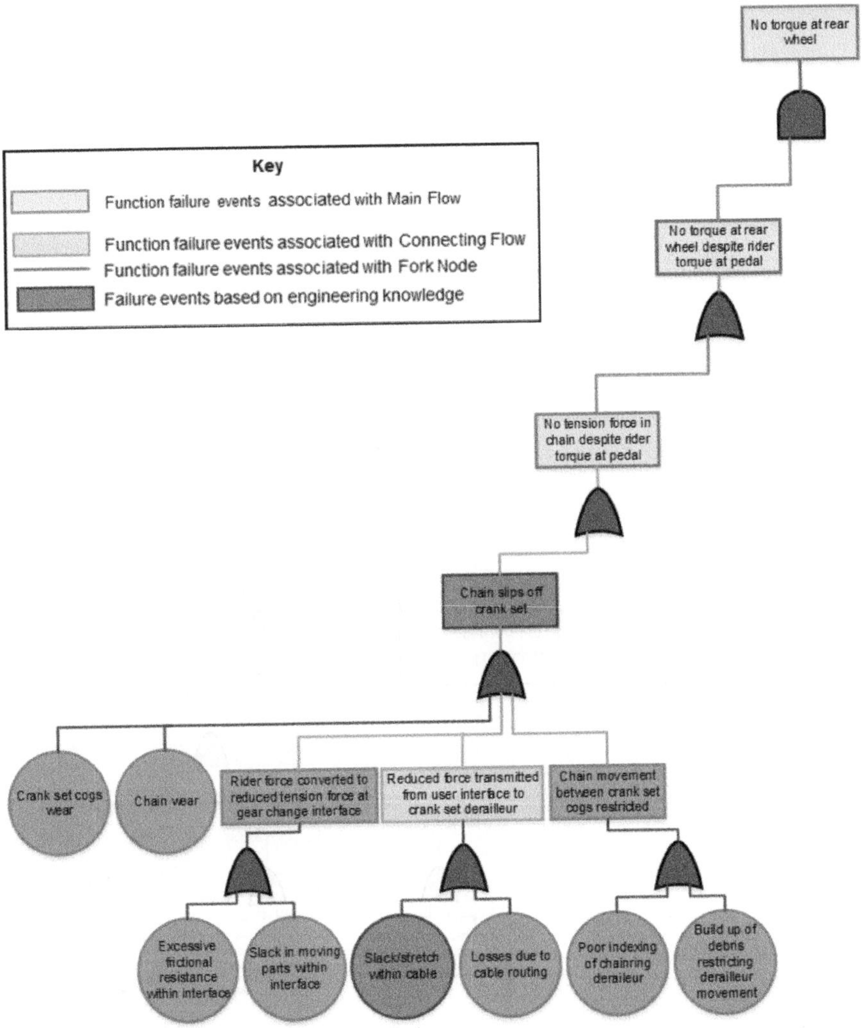

Figure 3.37. *Minimal cut set for human drive system section of FFT. For a color version of this figure, see www.iste.co.uk/elhami/mechatronics.zip*

3.6.4. *System of systems context for function fault tree analysis*

The eBike drive system FFT has been developed through a systems hierarchy starting by deriving the system-level FFT from the system-level SSFD and function tree. The subsystem and component levels of the tree were added to the system-level FFT having derived these from subsystem and component-level SSFDs and function trees. The eBike drive system is itself a system of the overall eBike, meaning that the drive system FFT can be seen as part of an overall eBike FFT. Such an FFT would be extremely large and cumbersome if developed to the same level of detail as the drive system FFT and as such would be better structured as a series of separate but interlinked FFTs. The development of interlinked SSFDs for systems, subsystems and components within a systems hierarchy lends itself to such a structure for the FFTs as Figure 3.38 illustrates.

The overall structure of the eBike SSFDs follows naturally from a functionally driven top-down product design process since the overall eBike must achieve the high-level functions of:

– store DC electrical power;

– generate AC electrical power;

– generate and transmit motive force;

– exert motive force;

– support and balance rider;

– decelerate the eBike;

– steer the eBike.

This chapter has focused on the bicycle level design solution that *generates and transmits motive force*, namely the drive system. The interlinked FFTs mirror the structure determined by the interlinked SSFDs both down the systems hierarchy and across system levels because of the relationship between the FFTs and the functional decomposition enabled by the associated SSFDs. Linkage between individual FFTs can be displayed on the FFT in the standard manner for fault trees using the triangular *transfer in/out* symbols (Veseley et al. 1981). It is a moot point whether some of the less complex systems, for example, that which *exerts motive force* (wheels and tires), would benefit from the development of an SSFD over a more standard engineering approach in achieving the functional decomposition although it should be recognized that the eBike is a significantly lower complexity system than say an electric car. Nevertheless, the relationship between the functional decomposition expressed in the form of a function tree and the corresponding FFT still applies.

The SSFDs and hence FFTs are partitioned on a functional rather than a hardware basis as several components on the eBike generate multiple high-level functions, for example, the handlebars serve both the functions of *support and balance rider* and *steer the eBike*. Decomposing the SSFDs and FFTs on a functional basis decouples such functions in the SSFD analysis which if partitioned by hardware would be unnecessarily complex with multiple inputs, outputs and SSFD heuristic flows.

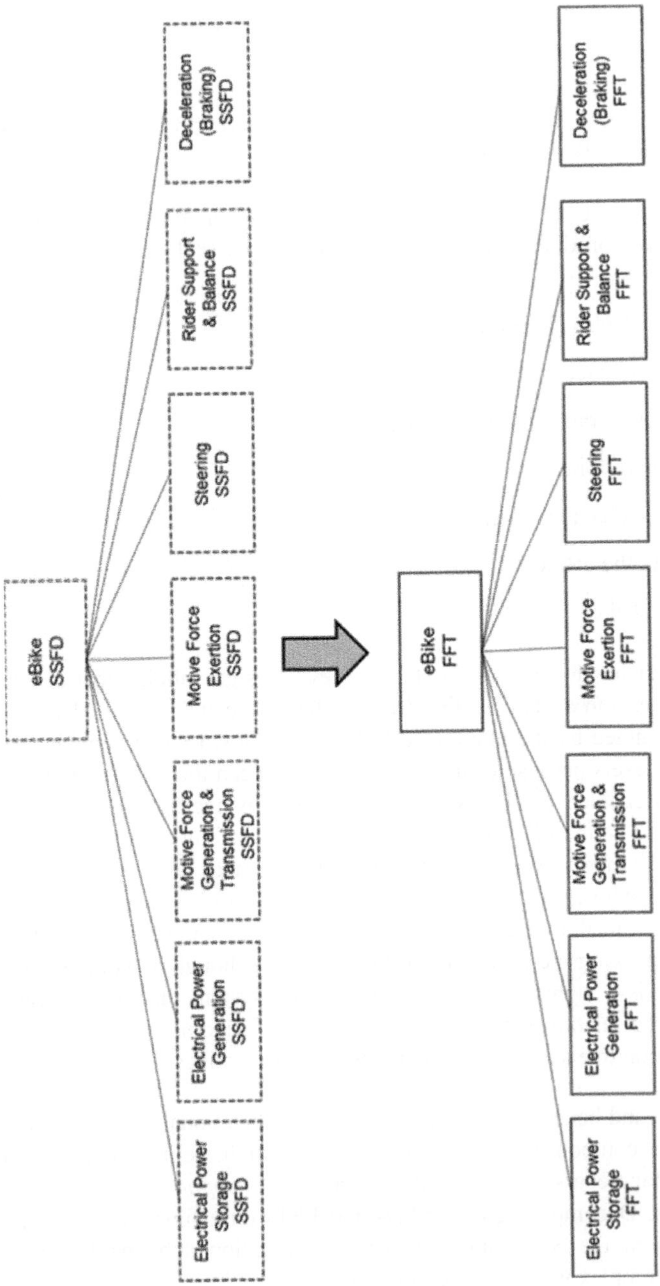

Figure 3.38. *eBike system of FFTAs based on the system of SSFDs*

The eBike drive system FFT of Figure 3.28 has three transfer-in symbols indicating linkage to the DC electrical power storage (charger and battery) and AC electrical generation (inverter) FFTs. The SSFDs, function trees and FFTs for these parts were developed by the eBike OEM working with the battery and inverter suppliers. While these two sets of documents were developed separately, it is important that they are analyzed in the context of both the higher-level system of the eBike and the subsystems with which they interact, for example, by considering them in the context of the electrical power main flow illustrated in the eBike drive system subsystem SSFD and FFT of Figures 3.12 and 3.15 respectively.

3.7. Relationship of FFTA to other engineering tools

As described in this chapter, the higher-level detail and structure of an FFT is derived from the corresponding function tree which itself is derived from an SSFD. This process is repeated over a number of system levels, illustrating the interrelated manner in which these engineering tools can be used within a systems engineering context. Complementary relationships exist between the FFT and other engineering tools, particularly FMEA and interface analysis. The documented outcome of the FMEA process differs from that of FFTA in that the FMEA is recorded in a tabular document while the FFT is graphical. The FFT and functional failure approach to the FMEA (AIAG/VDA 2019) are both driven by the identification of function failure modes, and in this sense, the SSFD and derived function tree can have equally important inputs to the development of both documents (Henshall et al. 2014). The FMEA includes significant information over and above that contained within an FFT including the effect, severity and probability of occurrence of a failure and any current and intended design action to eliminate or mitigate the effect of failure. The graphical nature of the FFT lends itself readily to the identification of potential failure mechanisms due to a combination of causes, while combinations of causes are more difficult to identify on an FMEA because each cause is documented within a line item and causes acting in combination can be several or even many line items apart. Given this, the FFT can act as a good counterpart to an FMEA particularly during the design of a new system, or in the use of new technology within an existing design, where potential failure mechanisms may not be well understood.

Interface analysis (IA) (Uddin et al. 2016) provides complementary information to the FFT by identifying the exchange of material, energy and information at system interfaces both between the internal elements of a system and between the internal elements and external entities. Exchange flows between internal system elements are identified on an SSFD while those between external entities, other than direct system inputs and outputs, are not and in this sense IA provides knowledge useful to the development of an FFT. Having identified an exchange, the effect of this exchange on system performance is determined and any functions required to reinforce those exchanges necessary for the system to operate as required or to mitigate the effect of exchanges that have a detrimental effect on system performance are established. Function failure modes associated with these functional requirements are the equivalent of the failure events documented on the FFT. IA is documented in tabular format with a supporting matrix containing higher-level detail. While, as with an FMEA, an FFT and its corresponding IA are likely to duplicate some information with IA containing more information overall than the FFT, the graphical nature of the FFT is its principle advantage.

The Reliability Block Diagram (RBD) (US DoD 1998) is a graphical representation of a system which can be used to identify combinations of causes and in this sense the RBD and FFT are comparable. The RBD differs from the FFT, however, in that its structural layout is based on knowledge of system architecture with each block representing a constituent part of a system. This can be compared to the FFT which, in being derived from a design solution neutral SSFD, means that an FFT can be developed earlier in the product design process before decisions on system architecture are finalized. The primary use of the RBD, like that of FTA, is for quantitative system reliability analysis with the two tools giving similar results given the same failure data (Benabid et al. 2018).

3.8. Discussion and conclusion

The methodology for developing an FFT as the mirror image of its corresponding function tree derived from an SSFD introduced by Campean and Henshall (2008) has been extended by subjecting the functional decomposition and the subsequent development of the FFT to the additional rigor gained by applying SSFD heuristics in the development of the SSFD. This rigor increases the structural integrity of the FFT so that it reflects the flow of materials, energy and information identified in the SSFD model of the system under investigation thus enabling subsequent analysis of the FFT to be consistent with the physics of operation of the system. This is particularly important when FTA is used as an integral part of the product design process. The top-down inductive nature of the development of an FFT means that the tree retains the structure dictated by the SSFD flow heuristics despite the fact that the majority, if not all, of the lower level basic events, may be identified through engineering knowledge and experience rather than being derived from a corresponding SSFD and function tree.

The inductive nature of the development of an FFT from an SSFD and function tree can be compared to the conventional, fundamentally deductive, development of fault trees where failure events are identified directly from historical failures and engineering knowledge, typically by asking how the top event can occur, hence identifying what Vesely (2002) terms *the necessary and sufficient immediate events* that cause the top event. This process progresses down the tree until the tree reaches a level of resolution corresponding to component failure. The authors' experience is that basing the development of a fault tree on, what is at best, an informal model of the system tends to result in causes being identified at different system levels at the same time which results in a tree that has no real separation of system levels, that is, multiple system-level failure events appear at any given level of the function tree. Developing fault trees based on an informal model of the system of interest runs the risk that the tree can wrongly reflect some aspects of the operation of the system. Where fault trees are developed on the basis of formal system models, such as in the automatic generation of fault trees, the examples quoted in the literature tend to be based on electronic circuitry or control loop applications.

In retaining a structure which is consistent with the physics of operation of the system being analyzed down to basic event level, the FFT facilitates the development of efficient and powerful countermeasures to potential failure during product design. Effort can be focused on intermediate events associated with flows of material, energy and information through the system. Optimizing such flows to be robust to the noise generated by those basic events which act as causal events to the intermediate events is an efficient way of managing potential system failures that the basic events might otherwise cause. The FFT also expediates the identification of basic events where the system failure they potentially cause is such that the development of countermeasures to failure falls outside of the scope of design optimization and a more radical approach is required, for example, functional redundancy, hardware redundancy, compensation or the choice of an alternative design concept. Equally importantly, along with the corresponding SSFD and function tree, the FFT facilitates an understanding of the consequences on system performance of adopting a particular design-led countermeasure in order to ensure that in preventing or mitigating the effect of one potential failure event the design countermeasure does not inadvertently cause other failure modes elsewhere in the system.

The FFT derived from an SSFD which has been developed using SSFD state flow heuristics is inherently based on a dynamic model of the system since the states in the SSFD incorporate the attributes of time and location. Thus, for example, when as a result of analyzing an FFT design optimization is conducted to develop countermeasures to failure events based on the use of a mathematically based model of the state flow of material, energy and information in a system, this analysis reflects the dynamic nature of the system. A system may operate in multiple different modes with some failure events being unique to a specific mode or not being feasible across all modes. In this situation, it should be relatively straightforward to adapt an FFT derived for one mode of operation to reflect an alternative mode thus making the identification of such failure events, and the subsequent incorporation of their effect on system performance into the optimization model, possible.

The ability to develop a series of SSFDs for a system on a top-down basis at different system levels translates directly to the structure and content of corresponding FFTs with the FFTs reflecting the increased granularity of failure events within a strict systems hierarchy based on the physics of operation of the system. What would otherwise be an extremely large and complex FFT for even a moderately complex system, can be developed top-down as a single system-level tree followed by a series of containable subsystem trees linked to the system-level tree. A further, more expansive, series of containable component-level trees (CLTs) can then be developed with each CLT linked to an immediately higher-level subsystem tree. The interlinked structure of the FFTs lends itself to different engineering teams developing FFTs that correspond to their area of expertise both within and across an OEM and part supplier companies. Developing a series of containable interlinked FFTs top-down in a system hierarchy is compatible with a systems approach to product design and development. The top-down inductive approach can be compared to the bottom-up deductive approach of developing individual component fault trees and subsequently linking them to form an overall system tree.

Setting the fault tree in a functional context also allows coherent two-way information flow between the FFT and other functionally driven product development tools additional to the SSFD and function tree such as FMEA and interface analysis. The underlying foundation of these tools is that their use is reliant on a functionally based understanding of the system being studied such as that provided by an SSFD. Such a foundation facilitates the design of a product with inherent safety and reliability and is key in using the FFT in a complementary manner to other tools as an integral part of the product design process

Despite the clarity provided by the systems approach to the development and analysis of FFT, the rigorous attention to detail facilitated by the understanding of the operation of the system being analyzed provided by an SSFD means that the use of FFT as a part of the product design process takes significant time and effort. Product design within a vehicle development program is usually rigidly time bound which makes it difficult to leverage the full power of the FTT and/or complementary engineering tools during the design process itself, potentially compromising the dependability of a design. One way of dealing with this challenge is to complete generic development and analysis outside of any vehicle program that can then be tuned to specific program needs. The design solution neutral nature of an SSFD and hence any function tree and FFT derived from it facilitates this strategy.

In conclusion, the use of the SSFD heuristics in developing an SSFD from which an FFT is derived significantly enhances the development and analysis of the FFT particularly when the FFT is used as an integral part of the product design process alongside other functionally based engineering tools. The enhanced FFT becomes a powerful way of identifying efficient and effective design-led countermeasures to potential failure system failure, thus helping to ensure the dependability of the product in customer usage.

3.9. References

AIAG/VDA. (2019). *Failure Mode and Effects Analysis (FMEA) Handbook.* AIAG, Southfield, MI.

Alshboul, B. and Petriu, D. (2018). Automatic derivation of fault tree models from SysML models for safety analysis. *Journal of Software Engineering and Applications*, 11, 204–222.

Andrews, J. and Henry, J. (1997). A computerized fault tree construction methodology. *Proceedings of the Institution of Mechanical Engineers*, Part E, 171–183.

Banach, R. and Bozzano, M. (2011). The mechanical generation of fault trees for reactive systems via retrenchment I: Combinational circuits. *Formal Aspects of Computing*, 25(4), 573–607.

Benabid, R., Merrouche, D., Bourenane, A., Alzbutas, (2018). Reliability assessment of redundant electrical power supply systems using FTA, RBD and Monte Carlo simulation methods. *CISTEM*, IEEE.

Campean, F. and Henshall, E. (2008). A function failure approach to fault tree analysis for automotive systems. SAE Technical Paper, 2008-01-0846.

Chai, F., Tang, Y., Pei, Y., Liang, P., Gao, H. (2016). Temperature field accurate modeling and cooling performance evaluation of direct-drive outer-rotor air-cooling in-wheel motor. *Energies*, 9, doi.org/10.3390/en9100818.

Cristea, G. and Constantinescu. (2017). A comparative critical study between FMEA and FTA risk analysis methods. *IOP Conf. Series: Materials Science and Engineering*, 252, doi:10.1088/1757-899X/252/1/012046.

Cui, J., Liew, L., Sabaliauskaite, G., Zhou, F. (2019). A review on safety failures, security attacks, and available countermeasures for autonomous vehicles. *Ad Hoc Networks*, 90, 101823.

Dori, D. (2002). *Object-Process Methodology; A Holistic Systems Paradigm*. Springer, Berlin.

Dugan, J., Bavuso, S., Boyd. (1992). Dynamic fault-tree models for fault-tolerant computer systems. *IEEE Transaction on Reliability*, 41, 363–377.

Gonschorek, T., Zeller, M., Hofig, K., Ortmeier, F. (2018). *Fault Trees vs. Component Fault Trees: An Empirical Study, Computer Safety, Reliability, and Security*. Springer International Publishing, Cham.

Haasl, D. (1965). Advanced concepts in fault tree analysis. *System Safety Symposium*, University of Washington and Boeing.

Henshall, E., Campean, I., Rutter, B. (2014). A systems approach to the development and use of FMEA in complex automotive applications. *SAE International Journal of Materials and Manufacturing*, 7(2), 280–290.

Hofig, K., Joanni, A., Zeller, M., Montrone, F., Rothfelder, M., Amarnath, R., Munk, P., Nordmann, A. (2018). Model-based reliability and safety: Reducing the complexity of safety analyses using component fault trees. *Reliability & Maintainability Symposium*, 64, doi: 10.1109/RAM.2018.8463058.

Kabir, S. (2017). An overview of fault tree analysis and its application in model based dependability analysis. *Expert Systems with Applications*, 77, 114–135.

Kaiser, B., Gramlich, C., Forster, M. (2007). State/event fault trees – A safety analysis model for software-controlled systems. *Reliability Engineering and System Safety*, 92, 1521–1537.

Lai, W. and Sung, C. (1998). Motion analysis of a bicycle rear derailleur during the shifting process. *Mechanical Machine Theory*, 33(4), 365–378.

Lee, W., Grosh, D., Tillman, F., Lie, C. (1985). Fault tree analysis, Methods, and applications – A review. *IEEE Transactions on Reliability*, R-34(3).

Li, Y., Huang, H., Liu, Y., Li, H. (2012). A new fault tree analysis method: Fuzzy dynamic fault tree analysis. *Maintenance and Reliability*, 14, 208–214.

Liggesmeyer, P. and Rothfield, M. (1998). Improving system reliability with automatic fault tree generation. *International Symposium on Fault-Tolerant Computing*, 90–99.

Nadjafi, M., Farsi, M.A., Zio, E., Mousavi, A. (2018). Fault trees analysis using expert opinion based on fuzzy-bathtub failure rates. *Quality Reliability Engineering International*, 34, 1142–1157.

Nadjafi, M. and Farsi M.A. (2021). Reliability analysis of system with timing functional dependency using fuzzy-bathtub failure rates. *International Journal of System Assurance Engineering and Management*, 12, 919–930.

Nonaka, I. (1991). The knowledge-creating company. *Harvard Business Review*, November–December.

Pahl, G., Beitz, W., Feldhusen, J., Grote, K.H. (2007). *Engineering Design: A Systematic Approach*. Springer, London.

Sharvia, S. and Papadopoulos, Y. (2014). Integrating model checking with HiP-HOPS in model-based safety analysis. *Reliability Engineering and System Safety*, 135, 64–80.

Sharvia, S., Kabir, S., Walker, M., Papadopoulos, Y. (2016). Model-based dependability analysis: State-of-the-art, challenges, and future outlook. *Software Quality Assurance*, 251–278, doi.org/10.1016/B978-0-12-802301-3.00012-0.

Smith, E.A. (2001). The role of tacit and explicit knowledge in the workplace. *Journal of Knowledge Management*, 5(4), 311–321.

Uddin, A., Campean, I., Khan, K. (2016). Development of an interface analysis template for system design analysis. *Proceedings of the 20th International Conference on Engineering Design*, Design Society.

Ulerich, N. and Poers, G. (1988). On-line hazard aversion and fault diagnosis in chemical processes: The digraph + fault-tree method. *IEEE Transactions on Reliability*, 37, 171–177.

US Department of Defense (1998). Electronic Reliability Handbook. US Department of Defense, Washington, DC.

Vesely, W. (2002). Fault Tree Handbook. NASA Office of Safety and Mission Assurance.

Vesely, W., Goldberg, F., Robrts, N., Haasl, D. (1981). *Fault Tree Handbook, Systems and Reliability Research*. Office of Nuclear Regulatory Research, U.S. Nuclear Regulatory Commission.

Wang, J. and Ruxton, T. (2007). A review of safety analysis methods applied to the design process. *Journal of Engineering Design*, 8(2), 131–152.

Yazdi, M. and Zarei, E. (2018). Uncertainty handling in the safety risk analysis: An integrated. *Journal of Failure Analysis and Prevention*, 18, 392–404.

Yi, X., Dhillon, B., Shi, J., Mu, H., Zhang (2017). A new reliability analysis method for vehicle systems based on goal-oriented methodology. *Journal of Automobile Engineering*, 231(8), 1066–1095.

Yildirim, U., Campean, F., Williams, H. (2017). Artificial intelligence for engineering design. *Analysis and Manufacturing*, 31, 413–435.

Zeng, L. and Li, Z. (2011). Control system of sensorless brushless DC motor based on TMS320F240. *I.J. Information Technology and Computer Science*, 5, 51–58.

Zhang, X. and Wang, Y. (2011). A novel position-sensorless control method for brushless DC motors. *Energy Conversion and Management*, 52, 1669–1676.

4

Reliability for a Mature Product From the Beginning of Its Useful Life: The Different Types of Tests and Their Impact on Product Reliability

The challenges with which the industry is confronted nowadays require consideration of innovative design, technologies and conditions of use. The reliability or maturation of such products could only be achieved experimentally, either by tests during the development or during the operational deployment. The risks associated with the second approach are unacceptable – to the ultimate users who demand a reliable product and to the supplier because of the detrimental image and associated costs unreliable products can cause.

This chapter presents recommended practices that represent the current best practices in the matter of building the product maturation during product development and highlighting the role of experimental testing in this maturation growth.

4.1. Introduction

The maturity of a product should be reached when its development has ended. However, the state-of-the-art technology in most industries reveals an imperfection of this maturation. Very often, this immaturity creates non-qualitative costs, especially for innovative product development. Essentially, the causes of this can be bound to deficiencies in:

Chapter written by Henri GRZESKOWIAK.

– the analysis of the product life profile (i.e. from leaving the factory to service withdrawal, including service mission and maintenance);

– the analysis of the technical specification at the system level and its hierarchical make-up at lower sub-system levels;

– the engineering of piece parts;

– the relevance of the design choices that take into account the real conditions of use;

– the elimination of the weak points of design and technologies;

– the validation of the technical functions (internal justification to the manufacturer) as in those of the service functions (justification in front of the client);

– the efficiency to disclose the latent defects in production.

The branches of industry which by necessity require us to have mature products, and especially those not to perform maintenance on the site, space industry sector of activity and the medical sector when patient's safety is paramount (pacemakers, etc.): the designs and technologies implemented are mature because they are based on validated technologies, with design margins, and consideration of field experience. The design of those products may also give preference to evolutionary designs rather than revolutionary design. Furthermore, redundancy and possibilities of reconfiguration may be incorporated into the design.

The space industry sector of activity has to put in-orbit complex reliable satellites required to operate reliably, with no maintenance, for many years. The approach to development is generally very conservative, de-rating systems and giving preference to evolutionary design. However, competition is changing these rules and even the space industry is using more revolutionary designs.

4.2. The product life profile

The product life profile is the chronological profile of the product, which starts when it leaves the factory and ends at service withdrawal. The chronological profile will encompass product use, missions and maintenance. A product life profile is characterized by its environmental factors (or constraints of employment in the wider sense), by the nature and severity (amplitude and duration) of the environment and the products' anticipated usage profile with associated performances. The environmental factors for which a given performance must be achieved constitute the normal environments for a product. Some situations can induce reversible damage for this performance level; such environmental conditions can be said to represent "limit" cases. In some instances, products may be required to withstand extreme situations, causing irreversible damage to some operational performance criteria, but without raising safety concerns. The corresponding environment can be said to be an extreme case.

It is important in a life profile analysis to consider the input of the customers and other stakeholders. This can be achieved by going to the customers' site and meeting the end users, field operators and the people involved in maintenance. This should take into account the operator's concerns in more stringent applications and in the consideration of environmental factors which may not be anticipated by the designer. Such conditions may include rain, sand and dust and the humidity arising from the periodicity of opening the containers or the periodicity of monitoring the manometers, indicating the gas proofness of the seals.

The life profile analysis should not neglect the transitory situations which are very often most significant. It is also necessary to consider the possible degradation in conditions of use which could shift from the normal or reversible domain into the limit or extreme domain. For example, putting a screwdriver in an unintended opening in order to secure the product in a particular phase of maintenance could induce the deterioration of an electronic circuit located in the heart of the opening. Also, materials may degrade due to repeated exposure to hot, cold or humid conditions.

4.3. The product technical specification

The product technical specification is a contractual document, established by the customer and given to the supplier. It expresses the needs of the customer in terms of requirements. The technical specification also gives the conditions and strategy of compliance validation.

The technical specification must: express what the customer wants from the product, how the product will be used, constraints for the design and production or the logistic support. The technical specification must be passed down through the product assembly tree to the lower level of assembly (the part, the coat of paint, etc., which will be specified by the corresponding technical specifications).

One difficulty is to echo the technical specification system into the lower levels of assembly. To help this process, models can be developed and used; moreover, if a design choice leads to a particular effect that has not been taken into account (e.g. a sensitive frequency beyond the specified domain, or effects induced from an external level of assembly), the technical specification of the product may have to be updated.

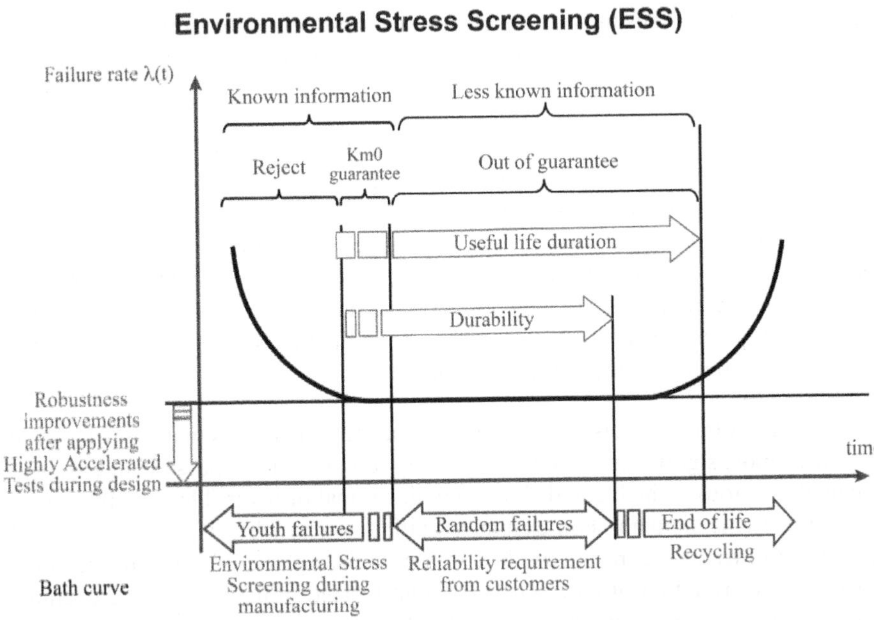

Figure 4.1. *Classical ESS, ESS mitigates the youth failures. For a color version of this figure, see www.iste.co.uk/elhami/mechatronics.zip*

The general requirements set out in technical specifications are related to either the functional performance or to the operational performance. The latter will determine the product dependability. Consideration of reliability is one part of the technical specification; it encompasses the ability of a material to produce the functions required in the given conditions of deployment and for a specified duration. However, this definition is not always perceived exactly in the same way by the end user of the material or by the contractor in charge of the development.

The end user has a perception that the overall reliability should integrate consideration of all the defects (deterministic or unpredictable) leading to a failure of the function. The developer has a perception that is not as critical of the material. In electronics, the general requirements in reliability concern the proportion of the defects, which have an unpredictable occurrence. It is this part that is frequently calculated from the standards on reliability such as MIL STD 217. An experimental verification may not be easy because of the unpredictable character of the defects. This may call for testing on a greater number of products in a brief time or on a more limited number of products during a time proportionally longer (10 times the expected MTBF in the case where a single product is tested). For mechanical systems, the reliability requirement is specified as the life duration of the product; it will generally be considered in a deterministic way throughout the product development and engineering. The verification test will usually be dedicated to the assessment of the life duration. It usually achieves this by applying an accelerated fatigue test, the level of acceleration being obtained by increasing the test level.

The reliability operational requirements in their current contents do not explicitly address the fact that the product is immature at the start of its development. The customer supposes that the state-of-the-art technology leads to a mature product from the beginning of its development.

4.4. The part (or component) engineering

The fast evolution of the components to market is a fundamental issue and the result of a reduction in the development cycle. A consequence of reducing the development cycle is the corresponding reduction in the number of windows of opportunity to improve maturity. The development costs of an emerging technology are growing and have to be shared between several companies and stakeholders.

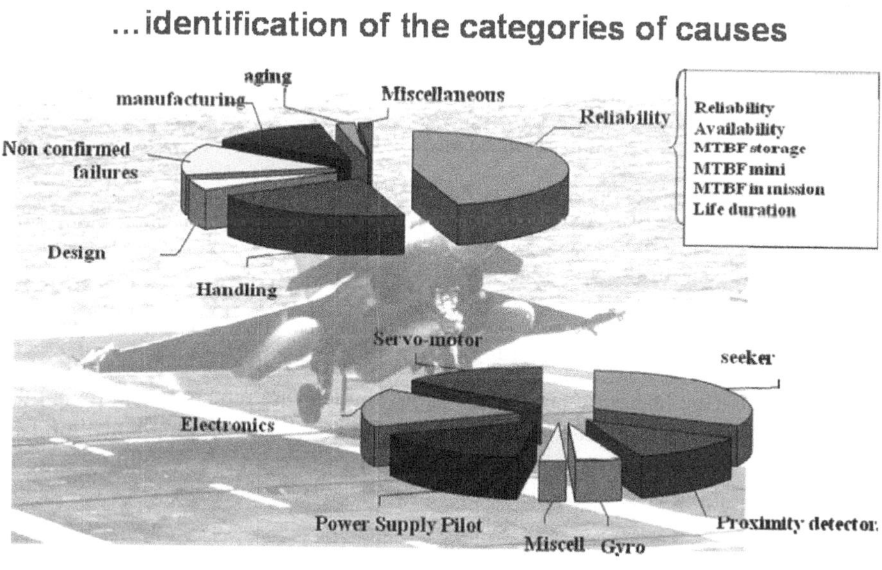

Figure 4.2. *Breakdown of origin of failures experienced on missile equipment. For a color version of this figure, see www.iste.co.uk/elhami/mechatronics.zip*

During the past five years, there have been marked technological evolutions which have resulted in the increased complexity of electronic circuits and the increased speed of operation, all at the cost of reduced design margins. There has been a decrease in the power supply voltages (5 V to 3.3 V and then 2.5 V), an increase of the thermal dissipation, and an increase in chip size within new circuit packaging. The continuous improvement in the manufacturing processes leads to a very low failure rate of the components and negates the need for final electrical tests. The qualified part list (QPL) approach is now being replaced by the qualified manufacture list (QML) approach; this puts the responsibility on the manufacturer to maintain their qualified manufacture list and to strive for continuous improvement.

The traditional approach for component life qualification and establishing component reliability are now considered inadequate for current products. The process of qualification must be reinvented.

From the viewpoint of the component manufacturer, it is necessary to:

– qualify the manufacturing processes by means of a test "vehicle";

– implement a computer-aided design with databases of manufacturing process reliability models;

– implement models of reliability for particular mechanisms such as electro-migration, injection of electrical charges simulating the thermal agitation, the loss of dielectric performance;

– use of the definite and complete models of simulation to confirm features;

– realize electric tests to characterize and evaluate the applications.

The manufacturer of components would also have to:

– administer the potential mechanisms of wear by maintaining the state of the art in design and by controlling the manufacturing processes;

– recognize that the interactions between the design and the reliability of the component should be considered during the design;

– administer the interactions between the manufacturing processes and reliability by a follow-up of the manufacturing processes. This will include implementing life duration tests to characterize the effects in the time, of evolutions of definition or of manufacturing processes.

From the viewpoint of the component user (prime, sub-contractor, etc.), a structure of engineering component support must:

– help the design teams in the choice of their components, of technologies, strategic circuits and suppliers, and thus from the beginning of the phase of preliminary design to evaluate test databases of the suppliers;

– carry out specific studies of the critical processes and components;

– generate follow-up sheets of performances;

– realize technological audits;

– assure that the parameters determining the objectives of long-term reliability are established and followed;

– assure that the rules of design and of component employment exist and are well applied and that the models of design and of manufacturing processes exist and are relevant.

The estimated equipment reliability is still widely based on the use of the constant rate of failure supplied by general standards (MIL HDBK 217 or RDF CNET 93 for the electronic components). This is despite numerous opinions which have attempted to put into perspective the validity of forecasts derived in this way. This is particularly significant for plastic components, which are still attached to the very pessimistic values (in the case of MIL HDBK 217) related to their problematic behavior at an extreme temperature (below 0°C and above 85°C). Conversely, for the use inside the temperature range 0–85°C, they can be as reliable as similar components in the metallic package.

A promising alternative school to this approach has been developing a body of information for several years by implementing the mortality rates and statistical laws. Supporters of this school consider that the defect of components arises from the deterministic consequence of the physicochemical phenomena, which are accessible by modeling. This school is represented by Maryland University (USA), which, for several years, has driven a software package entitled CALCE EPRC (Computer-Aided Life Cycle Engineering – Electronic Packaging Research Centre). CALCE EPRC's mission is to supply support to industry to assist it to develop (5–10 times) faster competitive electronic products. This support includes the supply of the methods of design and of manufacture, of techniques of simulation, of models, of experimental methods, of guides and of support for training.

Multidisciplinary researchers are driven to determine, estimate and update the methods and the practices for the design, manufacture, as well as the evaluation of reliable and competitive electronic products through studies in collaboration with the manufacturers. These studies, carefully chosen, encompass:

– failure component modeling;

– the accelerated tests;

– the methods of characterization;

– high-temperature components;

– software, modeling and automation;

– connectors and electrical contacts;

– modeling and characterization of the manufacturing processes;

– the optoelectronic circuits;

– the encapsulated plastic micro-circuits.

The technological innovations, which result from these studies, are the object of a transfer between the universities and the manufacturers. Research has led to very significant progress for the flexible printed circuits, the cards of printed circuits, the integrated hyper-frequency monolithic circuits, the plastic components, the mastery of metal migration, the optical and electric connectors, high density inter-connectors and the classical connectors.

Advanced models of thermo-mechanical fatigue were developed more particularly for surface-mounted technology (SMT) components.

Knowledge from research into the physics of failure, in material characterization and the effects of the temperature on the electronics is used to develop guides and procedures for accelerated tests. Furthermore, the analysis of finished elements is used to estimate the reliability of the advanced high technologies of interconnection and the latest packaging techniques. These research works were transformed successfully into industrial processes, into methods and recommended practices, given the concrete expression by the implementation of software on the industrial sites. Some of these companies have incorporated this software into their own tools.

Future developments will be aimed towards reducing the impact of the production costs while maintaining the quality and the reliability of the electronics. The domain of application covers the multi-chip modules (MCM), the electronics and opto-electronics at high temperature; a particular effort concerns the development of new electronic products.

For more information on the Calce program, see www.calce.umd.edu.

The Reliability Analysis Center (RAC) is a Department of Defence Information Analysis Centre (IAC). Since its inception in 1968, the RAC has served as a government and industry focal point for efforts to improve the reliability, maintainability and quality of manufactured components and items. To this end, the RAC collects, analyzes, archives in computerized databases and publishes data concerning the quality and reliability of equipment and systems as well as the microcircuit, discrete semiconductor and electromechanical and mechanical components. The RAC also evaluates and publishes information on engineering techniques and methods. Information is distributed through data compilations, application guides and software tools. Public and private training courses as well as consulting services are also provided. The US Air Force Research Laboratory provides technical management of the RAC.

For more information on the Reliability Analysis Centre, the reader can visit: www.rac.alionscience.com/InfoResources/Rac_ReliabilityData.html.

4.5. The performance-based requirement in design

Performance-based requirements of a product must be in terms of real conditions of use. The requirements should flow down from system to subsystem and component. Each item may be an existing evolution (i.e. no novelty of design, configuration or technology) or be new and innovative. Performance-based requirements can be based either on calculations or on validated models. Such calculations and tests need to take into account the variability of the product resistance to its environment by a strength/resistance approach.

Table 4.1 compares performance-based requirements with non-performance-based requirements.

Area of comparison	Performance-based requirements	Non-performance-based requirements
Purpose	Describe function products to be performed and level of performance	Describe how a product is to be designed and manufactured
Key criteria	Describe means for verifying performance	Describe means of ensuring that specified processes are followed
Design latitude, given to contractor	Allow contractor to determine the best ways to achieve results	Force the contractor to use prescribed methods and approaches
Responsibility	Responsibility for results clearly belongs to the contractor	Responsibility for results shared by the customer and contractor

Table 4.1. *Comparison of performance and non-performance-based requirements*

4.6. The elimination of weakness in design and technologies

One of the most effective recognized methods for revealing the weak points of design and/or technologies (and then create margins) is the application of aggravated tests. These are sometimes also called "highly accelerated tests" (see Figure 4.3). The principle consists of applying to the product step-by-step loadings (climatic, mechanical as vibration, electrical) and increasing the levels until the appearance of a failure. Then, failure analysis must be conducted to identify the appropriate corrective action (at the level of design, manufacturing process, condition of use). The testing then continues by increasing the levels of loading. The process terminates when the loading amplitudes are higher than the one resulting from the performance-based requirement. This loading should include any coefficients of conservatism necessary to address product resistance variability and strength variability.

Once a failure occurs, the technological analysis must identify if the failure is the consequence of a latent weakness or whether a technological limit has been reached. According to Hewlett Packard, the limit is reached when a small increase of the constraint produces a high increase of the number of failures.

The expected result of aggravated testing is to bring the product near to its inherent technological limits. Product maturity is then reached before the beginning of the production cycle. As a result, the operational performances and life duration margins are increased. The operational reliability is also improved and life cycle costs are reduced.

The number of units submitted to aggravated testing is dependent on the fraction of the population which is affected by a potential defect to be eliminated. As an example, if 2% of the population are affected by a considered defect and we want to precipitate it with a level of confidence of 95%, it will be necessary to test 145 units. On the contrary, if the whole population is affected, a single unit will be sufficient.

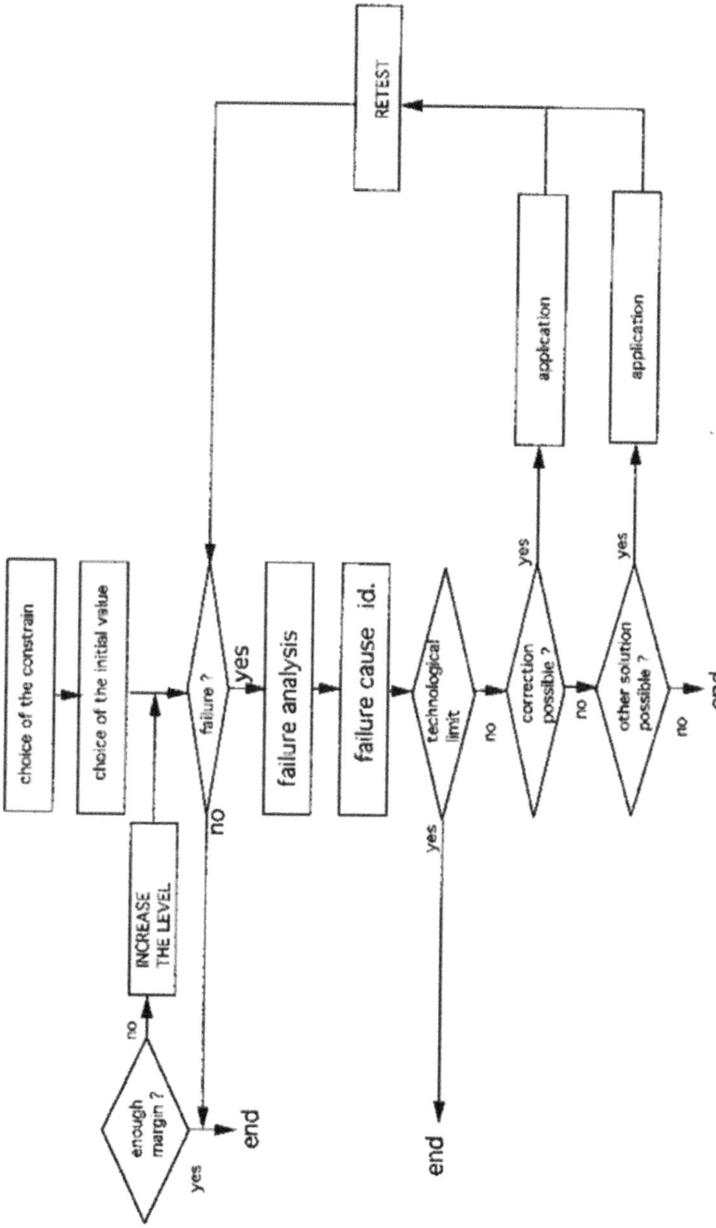

Figure 4.3. *Logical process of carrying out aggravated test process*

4.7. Uncertainty and test factors

Uncertainties and approximations can exist for various reasons. Typically, these can be grouped into environmental variability and equipment strength variability. The latter may be degraded as a result of product and material aging. These aspects are addressed in the following sections.

4.7.1. *Environment variability*

Environment variability arises as a consequence of the relative dispersion of the amplitudes of loading in the real environment. Variabilities exist in loadings from natural climatic (thermal and humidity) conditions as well as mechanical conditions such as vibration.

There are many sources of uncertainty in vibration environmental prediction. The main one is the variation of the vibration levels from one point to another one on complex structures, in particular at high frequencies. For a given situation, the real environment is defined by a single sample resulting from a short duration measurement; a factor CE can be applied to cover the possible dispersion on this value. Depending on the situation, CE can vary between 1.15 and 1.5 (CIN EG 1 and French Ministry of Defense 1999). In the case of road transport, to compensate for the uncertainties related to the variation of the characteristics of the vehicle, the terrain, the speed, the time of year, the driver and the measurement point, Holmgren suggests applying a factor of 1.15 to the overall rms value. In the case of the environment relative to a missile flight, a higher factor applies, equal to 1.4. If statistical results are used (frequency spectrum defined by a mean curve plus a few standard deviations), a factor of unity is applied.

4.7.2. *Equipment strength variability*

For reasons of cost, a single test is generally conducted. The selected specimen belongs to a population whose mechanical strength has a statistical distribution. A factor therefore needs to be applied to the real levels to make sure that the weakest item of equipment will be capable of withstanding the same environment. A factor of 1.15 is commonly employed for this purpose.

4.7.3. *Aging of equipment*

Aging of equipment is the alteration of strength over time (excluding fatigue and wear). Certain pieces of equipment are required to operate after a long storage period. To take this aging into account, a higher initial mechanical strength may be required by increasing the test levels by a certain factor, such as 1.5. If the test is conducted on equipment that is already aged, a factor of unity applies.

Testing generally requires conducting the vibration loading to be applied sequentially on each axis in turn. In most cases, the real vibration environment is an instantaneous vector of loading arising in the three orthogonal axes. In some cases, a multiplier of 1.3 is applied to take into account the fact that each component is necessarily less than or equal to the modulus of the resulting vector.

It has been observed that in the case of two resonances, which are not coupled but simultaneously excited in two different axes, a failure can occur in the real environment even though the product survived the test applied axis by axis. In such cases, a factor applied during the test does not necessarily solve the problem.

As will be realized from the preceding sections, there are many reasons for applying a conservatism factor. However, simply accumulating all the factors may lead to a large, non-realistic factor, as the conditions they represent never occur simultaneously. That is, the total conservatism coefficient is not derived by directly multiplying all the individual components.

This caution in determining the test specifications cannot be exaggerated. Lawrence, discussing vibration-isolating systems, considers that for structures, a factor of 2 often leads to a dimension and weight penalty of around 20%. However, in the case of random vibrations, a factor of 2 commonly leads to penalties of 100%. Nevertheless, such factors have been applied in the past. Generally, realistic total conservatism coefficients are:

– for acceptance tests: factors between 1 and 2 are applied to the real environment depending on the confidence in the data, with limited test duration to preserve the equipment but sufficiently long to allow functional testing;

– for qualification tests: increase of the acceptance test levels from 3 to 6 dB and of the test duration by a factor of 3–5, resulting in a test that is globally four times more severe and five times longer than the mission.

Any test will more accurately represent reality if an understanding of the real environment and material behavior is used instead of conservatism used out of "ignorance". This is possible, for example, by performing the tests on aged equipment, if necessary, or on three simultaneous axis test facilities if no axis is excited predominantly over the others in the field.

4.7.4. *The purpose of the uncertainty factor*

The environmental factor characterizing a given life profile situation will generate a different amplitude at each new occurrence. These amplitudes are randomly distributed, with a distribution law (in general normal or lognormal) and the associated statistical parameters: mean, variance (standard deviation), skewness and kurtosis. These parameters characterize the environmental factor variability.

The product resistance limit (characterizing the product subjected to the considered environmental factor) is also a random variable, representing the variability of the product to withstand environmental and imposed loading. In the absence of measurements, this variability can be characterized by considering an estimate of the corresponding coefficient of variation (i.e. the ratio of the resistance standard deviation to the resistance mean).

For each non-negligible environmental factor, it is possible to characterize the environmental factor variability and the corresponding material resistance variability.

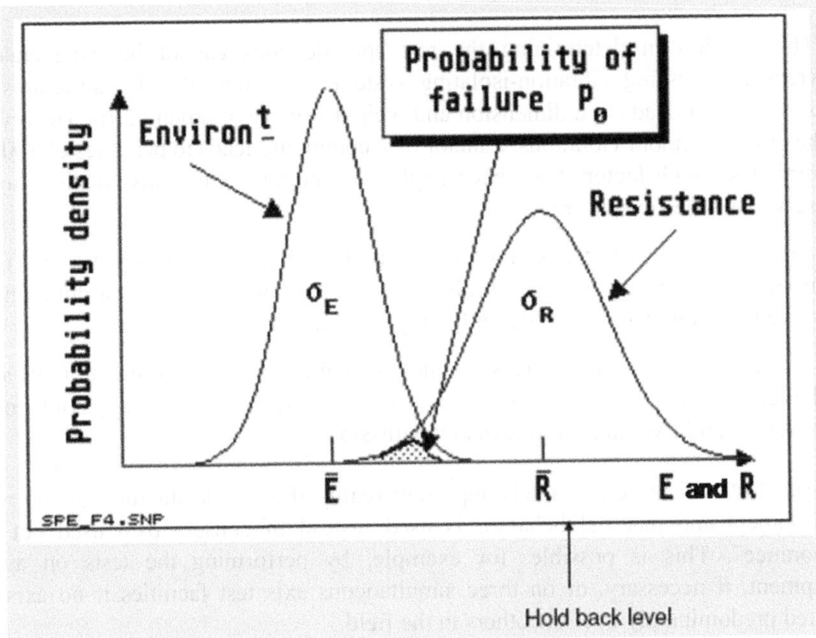

Figure 4.4. *Relationship between distributions in product strength and environment*

The existence of these two variabilities will establish a risk of failure (see Figure 4.4), indicated by the area of intersection between the two probability density curves respectively associated with the above two variabilities. The exact value of this risk can be calculated from the parameters characterizing the two probability density functions.

It will be noted that in the case of the material resistance variability, it is not normal to assume knowledge of the material mean resistance but only the ratio between the standard deviation to the mean resistance. This scalar value can be multiplied by the environmental factor mean value to permit the user to identify what the material mean resistance (to the considered environmental factor) required to meet the accepted risk of failure should be. This scalar is called the uncertainty factor. The result of this operation, which is still expressed in the same quantity as the original considered environment (or its representative), is referred to as the hold-back environment.

4.7.5. *The purpose of the test factor*

The aim of the test is to demonstrate that if the products under test pass the test, then the total deployed population in the field will disclose the rate of failure equal to the probability of failure taken for the calculation of the uncertainty factor (or safety factor). This would be simple if the population submitted to the test were infinite. In general, the items submitted are a small number (sometimes only one!). When the numbers of items submitted to the test are limited, the product strength (resistance) is located within an interval of confidence that becomes larger when the number of test items is reduced. The remaining part of this interval of confidence is the test factor.

In reality, it is not practical to undertake an infinite number of tests and generally only a few are used. When the number of tests is decreased, the interval of confidence, for a given probability, will become wider. To achieve the same guarantee of material resistance from a reduced number of tests, then the test severity must be increased by a new factor which is equal to the left part of the above interval of confidence. This new factor is usually referred to as the test factor. It is equal to one for an infinite number of tests and increases when the number of tests is reduced.

The above uncertainty and test factor approach supposes, for every performance sensitive to an environmental factor, the accepted probability of the corresponding failure. In the absence of data, the accepted probability is 10-3 when safety is not an issue; when safety is an issue then 10-6 is appropriate.

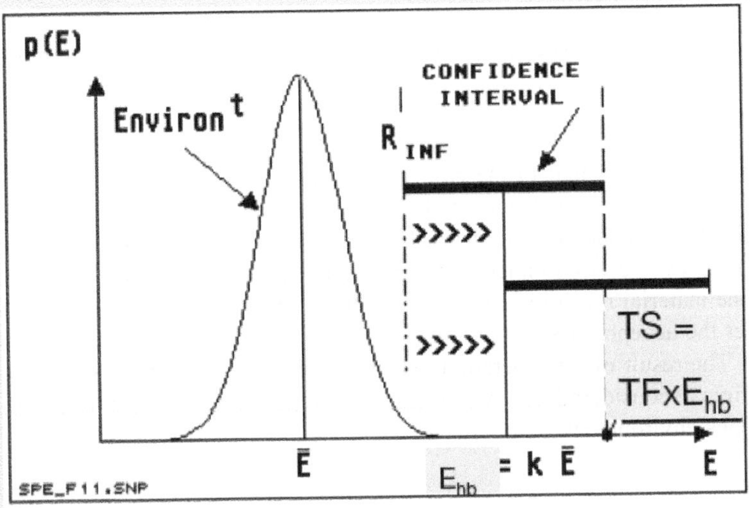

Figure 4.5. *Uncertainty interval*

4.8. Validation of the functions

The process of validation is a set of tasks where the engineer responsible for the product or the process recognizes that the product complies with the defined technical specification. This process leads to the validation of the functions and must be carried out in a progressive ascending way.

Whether the material is an evolution (no novelty of design, of configuration, of technology, of the conditions of employment) or a revolution (novelty of design, etc.), the validation will be based on calculations, models (validated by the previous return of experience) or on experimentation. The validation calculations and/or tests have to take into account the variabilities characterizing the component and/or material strength to resist the loading by an approach that will define the uncertainty factor and the test factor.

The actions of design and of validation can be carried out concurrently within the design phase. If the calculations and models used in the design phase to size a product are validated, and used within their validated area, then the corresponding action can be presented as an action of product validation.

The steps of the process leading to the validation plan consist of establishing a list of the tests and simulations, which are:

- optimizing it;
- building the logical chain of the withhold tests and simulations;
- defining the necessities to materials to test;
- confirming the global validation rationale at all the levels of assembly;
- defining the needs in test equipment;
- establishing tests and simulation work sheets;
- elaborating and optimizing the test plan;
- establishing, spreading and updating the test plan.

4.9. Environmental stress screening and HA-ESS

The principle of ESS consists of subjecting the units, during the production cycle, to various stimulations (vibrations, rapid variation of temperature, etc.) in order to precipitate latent defects inherent to the manufacturing processes, weak components and, in new products, weaknesses of design.

Mass production of equipment or highly integrated components (in general, mature components) do not require ESS but this is instead replaced by manufacturing process evaluation and control. ESS is essentially only a stop gap action, necessary as long as variability exists in the production processes. ESS can be superfluous for some types of components, such as the capacitors (except tantalum), resistors, diodes (generally), quartz, transistors. Some components may be screened either by the component supplier or the component user. Such components include some diodes, high power transistors, logic integrated circuits, tantalum capacitors and some relays.

ESS of electronic equipment has become essentially a classic strategy. The rate of defect precipitation for a product having a mature design and mature manufacturing processes generally ranges between 0 and 1%. Most of the manufacturers who implement ESS maintain it because they try to gain every tenth of a percent on the defects in the field, on the hypothesis that 10–50% of the precipitated defects during ESS would be disclosed in field deployment.

The application of ESS at the component level requires specific care as it can be expensive and is not without risk to the components. The choice of ESS program for a new component supposes an electrical and physical characterization, accelerated tests and a step stress approach. Abandoning an ESS approach may necessitate a compensatory component surveillance exercise.

For complex systems such as missiles, cycles of rapid change of temperature combined with vibrations constitute only two actions of typically nine different applications of ESS experienced by an electronic item before its integration at the full system level. It is only after this number of separate applications of ESS that the rate of defect stabilizes between 0 and 1%.

There are now more aggressive forms of ESS available which are known as HA-ESS (Highly Accelerated – Environmental Stress Screening) which is preceded by HAT (Highly Accelerated Testing), which tests the product limits for a state-of-the-art equipment in areas of design, technologies and product operation. The knowledge of these limits permits the application of more severe conditions than in classical ESS. However, care must be taken to prevent the consumption of a significant part of the products' potential useful life.

HAT has found a particular implementation known as HALT (Highly Accelerated Life Test), which applies several types of stimulation sets by different test equipment suppliers. Similarly, HA-ESS has also found a particular implementation, known as HASS (Highly Accelerated Stress Screening), which applies several types of stimulation sets by different test equipment suppliers.

The advantages put forward by users of the HASS method are:

– a simultaneous application of multi-axial random vibrations (typically generated by pneumatic hammers producing structural responses in 6 degrees of freedom);

– thermal cycling with a rapid change of temperature typically of 20°C/min to 60°C/min;

– functional equipment activation varying the power, the frequencies and the tensions;

– functional monitoring during the stimulations.

A relatively recent study "Environmental Stress Screening on 2000" was led by the National Centre for Manufacturing Sciences and the results were presented during the IEST Annual Meeting in 2001. The objective was to compare the cost/efficiency of various methods of ESS, either classical or more aggressive methods. The study reviewed:

– the approach referred to as MIL STD ESS (traditional approach based on NAVMAT 9492);

– the approach of HALT/HASS (Highly Accelerated Life Test/Highly Accelerated Stress Screen);

– the approach of ED/ACLT/ACSS (Electro-dynamic Accelerated Life Test/Accelerated Stress Screen); it should be mentioned that this is the same as HALT/HASS but using electro-dynamic shakers instead of electro-pneumatic shakers;

– ESS by thermal liquid/liquid shocks, LESS (Liquid Environmental Stress Screen).

The application of thermal shocks is often employed to precipitate latent defects in mechanical, electronic products and assemblies. The efficiency of this form of ESS loading is limited by the speed of the temperature variation. However, liquid/liquid thermal shocks are able to produce rates of 10,000°C/min. Nevertheless, considering that some components are not waterproof, the application may have limited application.

The results of the National Centre for Manufacturing Sciences study were:

– it allowed a better understanding of the failure mechanisms, easier choice of the screening method and its tailoring;

– the products that contain electronic component defects and then the liquid/liquid ESS will be adapted to precipitate them;

– on the other hand, the product contains more manufacturing defects and the HA-ESS approach is more effective;

– not one method among the above four is determined as the most efficient, irrespective of the product, the processes of manufacture or the packaging technologies used;

– military and commercial products can be subjected to HALT/HASS without any risk of damage.

For the products subjected to the liquid/liquid ESS, there were no defects bound to this method. The capacity of the product to be screened with this method depends on the materials of the product, which will not be damaged by the fluid medium.

4.10. Conclusion

The maturation of products can be achieved at the beginning of their exploitation. This corresponds and aligns to a legitimate and implicit user expectation. Methods exist to assist towards this objective, and many are presented in this document. The methods have in common that they assume ongoing efforts to improve maturity in all phases of development and manufacture of products. Experience indicates that to achieve a mature product, it is necessary to mobilize resources and numerous methodologies that are sometimes new.

4.11. Appendices

4.11.1. *Appendix 1: Types of tests*

4.11.1.1. *Accelerated tests*

During accelerated testing, a product is subjected to loading and conditions above those it would be expected to experience in service in order to accelerate the occurrence of failures but in reduced duration.

They are three types of accelerated tests:

Type 1: those based on acceleration by reducing the number of experiments, based on the design of experiments.

Type 2: those where the acceleration is based on statistical models.

Type 3: those where the acceleration is based on physical models of failure.

4.11.1.2. *Reliability growth tests*

Reliability growth is the improvement in the reliability of a product (component, subsystem, system) due to a well-structured process of finding reliability problems and monitoring the increase of the product's reliability through successive phases in time. A comprehensive reliability growth program is developed based on three important factors:

– management, where the decisions are made to keep the program moving;

– testing, where all the weaknesses and failure modes are found in the design and manufacturing process;

– failure identification, analysis and fix (FIAAF), where the cause of failure is isolated, analyzed and then fixed.

Reliability growth can be characterized after a small amount of tests and improvements (e.g. by the model by Duane).

4.11.1.3. *Highly accelerated tests*

A development test where the applied constraints have arisen in a progressive way in very upper values in comparison to the qualification specified values and the essential objective of which is to investigate the limits of functioning and of destruction of the product in order to push them away, by suitable actions, in the limits imposed by the state-of-the-art technology of its constituents. This is defined as the stress level at which a small increase in stress causes a large increase in the number of failures. Highly accelerated tests bring robustness or design margins to the product.

4.11.1.4. *Development tests*

Development tests are used to determine the aptitude of components, materials and/or designs to meet the requirements (functional and operational) of a given specification. These tests may prepare for the qualification, the reliability growth test (RGT) or the reliability demonstration tests. However, they can also contribute to prepare product development decisions.

4.11.1.5. *Qualification tests*

The qualification test is a series of tests realized on one or several examples of a product, to demonstrate compliance with the requirements (functional and operational) of a given specification.

4.11.1.6. *Reliability demonstrations*

A reliability demonstration test (RDT) is the demonstration of the reliability of a product. In particular, the RDT will generate data from which it will be possible to establish the average time before the product fails.

4.11.1.7. *ROSE robustness testing*

ROSE means robustness specification for environmental test. The test objective is to find weaknesses at an early design phase, particularly at the conceptual design and component specification phase. The key element is that components and products are tested until they fail in order to identify the relevant environmental test severity. With the knowledge of this severity, estimates can be made of the behavior, reliability and risks involved in using the design in-service.

4.11.1.8. *Robustness testing*

MEOST means multiple environmental overstress testing and is described by Bhote and Bhote (2004). The intention is to address multiple environments in terms of mutual influences and synergism. This is different from sequential testing, where tests are done in a certain order.

4.11.1.9. *Environmental Stress Screening (ESS)*

ESS is a test or a set of tests intended to remove defective items or those likely to exhibit early failures. The applied stress levels are normally similar to those expected in field operations.

ESS is often performed at the assembly level (PCB) or equipment level on a 100% basis to find defects and/or systematic failures during a preproduction phase or to provoke early failures during the series production and therefore to improve the reliability of a lot of items by elimination of the defective ones. To be efficient and cost-effective, ESS has to be tailored to the kind of item and the production process used.

4.11.1.10. *Highly Accelerated Life Testing (HALT)*

HALT is a commercial variant of highly accelerated test (HAT). Both are intended for the identification of design weakness by applying a series of environmental tests that typically include: thermal, vibration, power cycling and a combination of these environments. The thermal conditions can include very rapid variations of temperature (often with nitrogen cooling), while the vibrations are sometimes (but not only) applied in multi-axes by multi-input pneumatic hammer excitations.

During HAT and HALT, the item under test must be operational and its performance monitored. Contrarily to ESS, HALT is not a screening method but a test intended to verify the result of a development or the quality of the preproduction item. The results of HALT are used to define the stress limits for HA-ESS. Not surprisingly, the applied stress levels for HALT are higher than for HA-ESS. To reduce the test time for the items to be tested, the applied stresses are much higher than those expected in field operations. Indeed, the applied stresses may bear little or no relationship with those experienced in field conditions. For this reason, HALT is essentially a destructive test.

4.11.1.11. *Highly Accelerated Environmental Stress Screening (HA-ESS)*

HA-ESS is a screening method, like ESS, but the applied environmental severities are far above the field requirements while staying below the destructive limits. The aim of HA-ESS is to increase the reaction rate of the failure mechanisms and consequently to reduce test time.

The purpose of HA-ESS is not to highlight design problems, but to detect any defects in the unit that have arisen during the manufacturing/production process before the unit comes into use. The advantage of HA-ESS, compared to ESS, is partly the time factor as well as the ability to detect weak points which would not be detected by normal ESS. Typically, the test severities for HA-ESS are determined by the HAT results.

4.11.1.12. *Acceptance tests*

Acceptance testing is performed on a system (e.g. software, lots of manufactured mechanical parts, or batches of chemical products) or sub-system prior to its delivery. In some engineering sub-disciplines, it is known as functional testing, black-box testing, release acceptance, QA testing, application testing, confidence testing, final testing, validation testing, usability testing or factory acceptance testing. As the item has to go into service after the application of the severities, they are usually less than the qualification test severities, but set based upon consideration of the qualification test severities.

4.11.1.13. *Production Reliability Acceptance Tests (PRAT)*

The production reliability acceptance test (PRAT) is a process to demonstrate that the production standard hardware continues to meet the reliability requirements throughout production. This is normally achieved by subjecting a selected sample of production hardware to environmental conditions and functionally tested as specified in the PRAT plan that will be developed during pre-production.

4.11.2. *Appendix 2: Frequently asked questions on reliability test types*

4.11.2.1. *Accelerated tests*

4.11.2.1.1. What are the objectives of the test?

The prime objective of accelerated testing is to achieve faster investigations by reducing the test duration.

The principle of accelerated testing is that the product is subjected to conditions of use or to constraints of environment amplified in order to accelerate the mechanisms of failure and to reduce the necessary duration to estimate some behavioral characteristics of the product in the normal conditions of employment. Accelerated testing should not be confused with "highly accelerated testing" whose aims are very different. Accelerated testing has nothing to do with ESS or highly accelerated testing.

In order to successfully apply an accelerated testing strategy, certain conditions of application must be adhered to:

– the relationship between the mechanism of failure with the amplitude and duration of the applied conditions and the parameters involved;

– the induced mechanisms of failure have to be representative of those arising in the normal conditions of use;

– synergism and interactions between several constraints of use and of the applied environment must be taken into account.

Testing is done until failures take place, and the key is to understand the failures and their relationship to the applied stresses and to correct them. Accelerated life testing is performed to obtain information about the product (component, module, system, etc.) lifetime distribution or a particular reliability parameter in a timely manner. This can be done in two different ways or in a combination of them:

– using time compression, for example, by changing the length of on and off cycles;

– accelerating the loading conditions.

The purpose of life testing is to identify the relevant failure mechanisms that would occur, and correlate these with the point in the product's life the failure would occur. All the stresses during the product's whole life cycle need to be addressed, for example, transportation, storage and the actual use conditions. These conditions usually include different and changing temperatures, humidity, mechanical stresses (vibrations, shock, bumps, etc.), electrical and electromagnetic phenomena, etc.

4.11.2.1.2. What are the drivers for choosing the test?

The need to shorten time to market is the main reason for adopting an accelerated testing strategy.

4.11.2.1.3. What level of assembly of the product is normally submitted to test?

Quite often, the greatest benefit of accelerated tests (in terms of knowledge acquisition) is achieved by performing tests at the lowest level of assembly. Tests should also be done early in the product development process. However, testing should be continued during the different phases of the product development and the production phases.

4.11.2.1.4. How many items are normally submitted to test?

For statistical confidence, the need is often higher than what is practically possible; about five being a minimum but several dozen would be needed.

In general, one; the application on a higher amount might be a response for product variability considerations: see hereinafter.

4.11.2.1.5. What type of product is appropriate for this test type?

All types of products may be submitted to accelerated testing

4.11.2.1.6. What is the typical duration of a test program?

Accelerated stress testing can be done variably in days or a few weeks. Life testing typically requires several weeks, even several months.

4.11.2.1.7. Should separate or combined environments be considered?

Different types of stresses are needed to precipitate the different failure mechanisms. Their simultaneous combination is sometimes necessary. A proper combination of stresses may be used to realize synergistic effects – for example, thermal cycling and vibration. Humidity is also usually applied with constant or varying temperature.

4.11.2.1.8. Are variations in product strength and environmental stresses considered?

The accelerated test can take into account variations of product strength.

4.11.2.1.9. Does the test supply knowledge on reliability parameters?

Life testing aims at getting an estimate of, for example, the failure rate (or MTTF, MTBF) or the onset of wearout mechanisms (lifetime) or at least the relevant failure mechanisms.

4.11.2.1.10. Does the test consume product life?

Accelerated stress tests and life tests are destructive.

4.11.2.1.11. What are the main norms, standards, technical references relevant to this test type?

Additional relevant information can be found in:

– IEC 60068-2 Environmental testing, accelerated stress tests;

– IEC 60749 Mechanical and climatic test methods (semiconductor devices), MIL STD 810 G;

– STANAG 4370;

– DEF STAN 00-35;

– NF 50-144-1 to 6.

4.11.2.2. *Reliability growth tests*

4.11.2.2.1. What are the objectives of the test?

Reliability growth testing is performed to assess current reliability, identify and eliminate faults, and forecast future reliability. The reliability figures are compared with intermediate reliability objectives to measure progress so that resources can be directed to achieve the reliability goals in a timely and cost-effective manner. Whenever a failure occurs, corrective action is undertaken to remove the cause. For hardware, growth testing is the process of testing the equipment under both natural and induced environmental conditions to discover latent failure modes and mechanisms to ensure that all performance, design and environmental problems have been resolved.

Reliability growth is the improvement in the reliability of a product (i.e. component, subsystem or system) or service over a period of time due to changes in the product design and/or the manufacturing process. Reliability growth analysis (RGA) concerns itself with the quantification and assessment of parameters (or metrics) relating to the item's reliability growth over time. Reliability growth management concerns itself with the planning and management for the item's reliability growth as a function of time and resources.

Reliability growth is a quantitative methodology; it is used to analyze the improvements made on a product design and verify whether a certain reliability goal has been reached. It is usually performed in normal conditions (unlike HALT which is performed in highly stressed conditions). In general, the first prototypes produced during the development of a new complex system will contain design, manufacturing and/or engineering deficiencies. Because of these deficiencies, the initial reliability of the prototypes may be below the system's reliability goal or requirement. In order to identify and correct these deficiencies, the prototypes are often subjected to a rigorous testing program. During testing, problem areas are identified and appropriate corrective actions (or redesigns) are taken. Reliability growth is the improvement in the reliability of a product (component, subsystem or system) over a period of time due to changes in the product's design and/or the manufacturing process. A reliability growth program is a well-structured process of finding reliability problems by testing, incorporating corrective actions and monitoring the increase of the product's reliability throughout the test phases. For more info about reliability growth, refer to: http://www.weibull.com/relgrowthwebcontents.htm#Introduction.

4.11.2.2.2. What are the drivers for choosing the test?

Reliability growth testing is performed to assess current reliability, identify and eliminate faults, and forecast future reliability.

4.11.2.2.3. What level of assembly of the product is normally submitted to test?

In general, the first prototypes produced during the development of a new complex system will contain design, manufacturing and/or engineering deficiencies.

4.11.2.2.4. How many items are normally submitted to test?

Usually several items, typically between one and ten.

4.11.2.2.5. What type of product is appropriate for this test type?

Typically, innovative mass volume products which can be revolutionary either in design, technologies used or conditions of usage. Also, potentially sensitive products or those with safety implications.

4.11.2.2.6. What is the typical duration of a test program?

The typical duration of testing is measured in weeks.

4.11.2.2.7. Should separate or combined environments be considered?

This is mostly a combined test sequence.

4.11.2.2.8. Are variations in product strength and environmental stresses considered?

Product strength and environment stress variabilities are not considered.

4.11.2.2.9. Does the test supply knowledge on reliability parameters?

Reliability information generated and recorded over time can be used to observe trends in the reliability of the product. The term "growth" is always used since you assume that the reliability of the product will increase over time as design changes and fixes are implemented. In other words, reliability growth is a projection of the reliability of a system, component, unit (or service) to some future development time. This projection is based upon information currently available from predictions or prior experience on identical or similar systems. Monitoring the reliability, the mean time between failures (MTBF) and the failure rate of the system, equipment or product also establishes a trend in the increase in the reliability, the increase in the MTBF or the decrease in the failure rate. This is achieved with engineering, research, development, test-analyze-and-fix (TAAF) and/or test-analyze-and-redesign (TAAR) procedures until it passes its acceptance tests and/or is delivered to the end user.

4.11.2.2.10. Does the test consume product life?

Product life is consumed by this test.

4.11.2.2.11. What are the main norms, standards, technical references relevant to this test type?

MIL-HDBK-189 Reliability Growth Management, February 1981, 155 pages. This handbook provides procuring activities and development contractors with an understanding of the concepts and principles of reliability growth, advantages of managing reliability growth, and guidelines and procedures to be used in managing reliability growth. It should be noted that this handbook is not intended to serve as a reliability growth plan to be applied to a program without any tailoring. This handbook, when used in conjunction with knowledge of the system and its development program, will allow the development of a reliability growth management plan that will aid in developing a final system that meets its requirements and lowers the life cycle cost of the fielded systems (this document describes the Duane method of reliability growth and which became the Duane AMSAA methodology which today is described as the Crow-AMSAA reliability growth model).

4.11.2.3. *Highly accelerated tests*

4.11.2.3.1. What are the objectives of the test?

Tests where the applied constraints have risen in a progressive way in very upper values in comparison to the qualification specified values, the essential objective of which is to investigate the limits of functioning and of destruction of the product in order to push them away, by suitable actions, in the limits imposed by the state-of-the-art technology of its constituents.

4.11.2.3.2. What are the drivers for choosing the test?

Reasons for using this test strategy include:

– time to market of the product;

– safety consideration (avionics);

– maintainability costs;

– technological limit: the limit of resistance imposed by the technology of a product or of a particular constituent, towards a given constraint (temperature, vibration, electric tension, etc.). This limit establishes so an unbridgeable barrier. Ex: melting point of a plastic building material, temperature, the maximum of a ***functioning a semiconductor in the silicon, elastic limit of a building material, etc.***

4.11.2.3.3. What level of assembly of the product normally submitted to test?

This type of test is normally undertaken on items at the lowest possible level of assembly.

4.11.2.3.4. How many items are normally submitted to test?

The statistical aspects in relation with the number of realized highly accelerated tests on a new product intervene to at least two levels:

– the capacity of the highly accelerated tests to discover relevant causes of failing;

– obtaining the limits of functioning and/or of destruction and the associated variableness.

With regard to the capacity of the highly accelerated tests to discover relevant causes of failing, two cases are considered:

– First case: the relevant causes of failing presumed are of abstract nature and have for effect to reveal a problem on every copy subjected to attempt, since an applied constraint exceeds the specified value. It is translated generally by an incapacity of margins, typical case of the use of a badly calibrated constituent. In that case, the realized highly accelerated tests on a very limited sample of copies, even on the single copy, turns out effective to reveal the existence of the relevant cause of defect and power so to correct it.

– Second case: the relevant causes of failing presumed can affect a priori only a weak fraction of the complete population of the product (e.g. less than 5% of the population). In that case, it is not very probable that a limited number of highly accelerated tests allows us to reveal this relevant cause. As example, the use of the model binomial indicates us that if we hope to reveal a relevant cause of failing affecting only 5% of the population with a probability of the success of 80%, it would be necessary to subject about 30 copies to the test of highly accelerated tests. Such quantity is totally incompatible with the imperative manufacturers. Consequently, the highly accelerated tests realized in the phase of conception/development on an always very limited number of copies cannot have a great efficiency towards the revealing of this type of defects. Only, highly accelerated ESS (HA-ESS) in the phase of production will be susceptible to reveal, in addition to the manufacturing defects, these defects of the conception, which would affect only a weak fraction of the products during their profile of life.

With regard to obtaining the limits of functioning and/or of destruction of the product, the major difficulty places itself in the statistical character associated with these sizes. So, obtaining in a different way the given value on the limit of functioning the single copy does not inevitably reveal the central value of the statistical distribution of this limit which could be associated with an important population of the tested product. In particular, this unit experimental value can be a value of the tail of distribution, placed to the left or to the right of the distribution. It proves why it is hardly recommended, when it is compatible with the imperatives of development, to make the highly accelerated tests on several copies and not on the single unity. The number of available copies always being very limited in the phase of development.

4.11.2.3.5. What type of product is appropriate for this test type?

The majority of products are appropriate for this test type. For some products, highly accelerated testing may induce failures that do not occur during the actual use. In such cases, for some product types, it may not be possible to rectify identified failures and meet other design and cost requirements.

4.11.2.3.6. What is the typical duration of a test program?

The duration of a highly accelerated test campaign is very brief: of the order of five days on average.

4.11.2.3.7. Should separate or combined environments be considered?

Combined environments are not normally employed but can be necessary to disclose a failure. For example, we can gradually increase the vibration level for a given rate of temperature variation or by gradually increasing this rate for a given level of vibration.

4.11.2.3.8. Are variations in product strength and environmental stresses considered?

The test does not address variations in product strength and environmental stresses.

4.11.2.3.9. Does the test supply knowledge on reliability parameters?

The test provides no evidence on reliability parameters.

4.11.2.3.10. Does the test consume product life?

Yes, the product under test is normally destroyed.

4.11.2.3.11. What are the main norms, standards, technical references relevant to this test type?

None currently.

4.11.2.4. *Development test*

4.11.2.4.1. What are the objectives of the test?

Development tests are used to determine the aptitude of components, materials and/or designs to meet the requirements (functional and operational) of a given specification. These tests may prepare for the qualification, or the RDGT or reliability demonstration, but not only this: they also can contribute to prepare R&D decisions.

Moving from the specified environmental requirements relating to operation, transportation and storage, the appropriate test procedures, the stress levels and the test duration will be chosen, considering the experiences with earlier development and the know-how of the test specialists.

The test results are used to make design decisions as well as to evaluate the "reliability" and/or the "lifetime" of a product and to prove the rightness of a design. Therefore, the test results, methods, tools and conditions will be documented and stored in such a way that the probability of losing or distracting is then very low.

4.11.2.4.2. What are the drivers for choosing the test?

The tests are done during the development and design phase of a project with two aims:

– to determine the ability of materials, components and/or designs to resist on effects of specified environmental conditions, where experience is missing and/or calculation or virtual simulation is not possible or not accurate enough;

– to determine the ability of a design solution to meet the functional requirements of a given specification. While being a substantial part of development costs and an important design tool, it is necessary to plan and to budget for this at the very beginning of a development project. The principal input for this planning activity is the product specification.

4.11.2.4.3. What level of assembly of the product normally submitted to test?

The test may be used on material, components and sub-assemblies.

4.11.2.4.4. How many items are normally submitted to test?

In general, one. But several may be retained for qualification in order to reduce the duration of it: the sequence of tests applied to each of them must be in relation with the real sequence expected in the field.

4.11.2.4.5. What type of product is appropriate for this test type?

Not relevant for those constraints covered by other specified tests (which must be stated in the test plan) such as highly accelerated tests.

4.11.2.4.6. What is the typical duration of a test program?

The duration of the test program will depend on the goal of the specific development test and the type of test: endurance (several hours or days), reliability (in general, 10 times the specified MTBF for one equipment or 10/n for n equipments).

4.11.2.4.7. Should separate or combined environments be considered?

As much as possible, the conditions should be separated. This may not be possible in some specific cases such as the presence of polymer material in a suspension (requires vibration and temperature).

To allow us to determine the effect of single stress factors on a material or item respectively to avoid the reciprocal influence of several stresses, the test conditions are usually special lab conditions, not equal to the real environmental conditions of the product.

Combined tests (humidity/temperature; vibration/temperature, etc.) are also often used, but normally applied on the final product and not for the purpose stated.

4.11.2.4.8. Are variations in product strength and environmental stresses considered?

Variations in product strength and environmental stresses should be considered. The uncertainty (or safety) factor and test factor are the best way to manage these variabilities.

4.11.2.4.9. Does the test supply knowledge on reliability parameters?

Yes, if the test is dedicated to reliability growth.

4.11.2.4.10. Does the test consume product life?

The sub-assemblies, components, submitted to test are not intended for delivery for the customer. They might be used for mock-up, study or prototype purposes.

4.11.2.4.11. What are the main norms, standards, technical references relevant to this test type?

The results of these tests are normally only for internal use (in general, not given to the customer). So, the reference to the standards is not mandatory. No known standards are related to the development tests.

4.11.2.5. *Qualification tests*

4.11.2.5.1. What are the objectives of the test?

Qualification testing is a series of tests realized on one or several examples of a product, complying with the same definition file, in order to assure that it meets all of its design specifications. This is also referred to as:

– Verification and validation (V&V) testing, or design approval. A series of tests conducted by the government procuring activity, or an agent thereof, to determine conformance of materials, or material systems, to the requirements of a specification which normally results in a qualified products list (QPL) under the specification. Qualification under a specification generally requires conformance to all tests in the specification. It may, however, be limited to conformance to a specific type, or class, or both.

Product certification or product qualification is the process of certifying that a certain product has passed performance and/or quality assurance tests or qualification requirements stipulated in regulations such as a building code and nationally accredited test standards, or that it complies with a set of regulations governing quality and/or minimum performance requirements. A formally defined series of tests by which the functional, environmental and reliability performance of a component or system may be evaluated in order to satisfy the engineer, contractor or owner as to its satisfactory design and construction prior to the final approval and acceptance.

4.11.2.5.2. What are the drivers for choosing the test?

As soon as the first prototype is built, the product must begin testing to assure that it meets all of its specifications (verification). Once this is complete, it must then undergo further testing to assure that it works as the customer intends to use the product (validation).

4.11.2.5.3. What level of assembly of the product is normally submitted to test?

Highest product level of assembly if the corresponding facilities exist and if the cost is reasonable. Otherwise, split in sub-assemblies that are qualified separately.

4.11.2.5.4. How many items are normally submitted to test?

One or several, but generally less than five.

4.11.2.5.5. What type of product is appropriate for this test type?

All the products should be qualified.

4.11.2.5.6. What is the typical duration of a test program?

In general, weeks for a whole system.

4.11.2.5.7. Should separate or combined environments be considered?

Separate environments. In some cases, combined environments should be used: for example, in the case of elastomeric suspension where vibration and temperature should be combined.

4.11.2.5.8. Are variations in product strength and environmental stresses considered?

Variations in product strength and environmental stresses are considered.

4.11.2.5.9. Does the test supply knowledge on reliability parameters?

The test does not supply knowledge on reliability parameters.

4.11.2.5.10. Does the test consume product life?

The test consumes product life.

4.11.2.5.11. What are the main norms, standards, technical references relevant to this test type?

– MIL STD 810 G STANAG 4370;

– DEF STAN 0035;

– NF 50-144-1 to 6.

4.11.2.6. *Reliability demonstration*

4.11.2.6.1. What are the objectives of the test?

A reliability demonstration test (RDT) is the process of demonstrating the reliability of a product. An RDT will be able to define in particular the average time before the product fails. We can run the RDT in an accelerated manner to obtain this information in order to decrease the development time.

There are in general three stages to a reliability demonstration:

– test – planning;

– testing/monitoring;

– final analysis.

Planning the test stage. In the planning stage, a reliability goal is established. Then, a reliability demonstration test model is chosen. We will then develop a decision matrix and outline all of the parameters that we must decide on prior to writing the plan, including types of stresses, number of units, length of test, confidence, etc., along with advantages and disadvantages for each. From this matrix, we will jointly decide on all of the parameters that will go into the RDT plan.

Testing and monitoring stage. In the testing and monitoring stage, the reliability demonstration test is performed and the results are monitored to determine the status compared with the goal. The test will be set up so that the monitoring will only need to be performed a few times a week, once a day at most.

Final analysis stage. In the final analysis stage, the data is compiled and a report is written on the results. The report will compare the achieved results to the goal and will show how the data was calculated. It will contain detailed information about the models used.

Frequently, a manufacturer will have to demonstrate that a certain product has met a goal of a certain reliability at a given time with a specific confidence. Often, it will be desirable to demonstrate that this goal has been met with a zero-failure test. In order to design and conduct such a test, something about the behavior of the product will need to be known, that is, the shape parameter of the product's life distribution. Beyond this, nothing more about the test is known, and usually the engineer designing the test will have to study the financial trade-offs between the number of units and amount of test time needed to demonstrate the desired goal. In cases like this, it is useful to have a "carpet plot" that shows the possibilities of how a certain specification can be met.

This methodology requires the use of the cumulative binomial distribution in addition to the assumed distribution of the products' lifetimes. Not only does the life distribution of the products need to be assumed beforehand, but a reasonable assumption of the distribution's shape parameter must be provided as well (Note: the exception to this is the exponential distribution, which does not have a shape parameter. This is equivalent to assuming a Weibull distribution with $\beta = 1$). Additional information that must be supplied includes the reliability to be demonstrated, the confidence level at which the demonstration takes place, the acceptable number of failures and either the number of available units or the amount of available test time. The output of this analysis can be the amount of time required to test the available units or the required number of units that need to be tested during the available test time.

Frequently, the entire purpose of designing a test with few or no failures is to demonstrate a certain reliability, RDEMO, at a certain time. With the exception of the exponential distribution (and ignoring the location parameter for the time being), this reliability is going to be a function of time, a shape parameter and a scale parameter.

$$RDEMO = g(tDEMO, \theta, \phi) \qquad (3)$$

where:

tDEMO is the time at which the demonstrated reliability is specified;

θ is the shape parameter;

ϕ is the scale parameter.

Since the required inputs to the process include RDEMO, tDEMO and θ, the value of the scale parameter can be "backed out" of the reliability equation of the assumed distribution, and will be used in the calculation of another reliability value, RTEST, which is the reliability that is going to be incorporated in the actual test calculation. How this calculation is performed depends on whether we are attempting to solve for the number of units to be tested in an available amount of time, or attempting to find how long to test an available number of test units.

Designing reliability demonstration tests is an integral part of any reliability program. These tests are often required to demonstrate customer reliability and confidence requirements. While it is desirable to be able to test a large population of units to failure in order to obtain information on a product's or design's reliability, time and resource constraints sometimes make this impossible. In cases such as these, a test can be run on a specified number of units, or for a specified amount of time, that will demonstrate that the product has met or exceeded a given reliability at a given confidence level. In the final analysis, the actual reliability of the units will of course remain unknown, but the reliability engineer will be able to state that certain specifications have been met. In addition to analytical methods (such as the parametric binomial, the non-parametric binomial and the exponential chi-squared methods), a simulation method is also possible.

4.11.2.6.2. What are the drivers for choosing the test?

Reliability demonstration tests have a tremendous value in many areas. They can:

– predict the life of the product in the field;

– determine whether the product is ready to ship.

The RDT is based on a combined vibration/thermal stress environment, usually for a single stress cycle profile chosen to be typical of in-service. The actual stress level may be accelerated above "typical" in order to gain a corresponding reduction in test time. The eventual test profile and levels selected for the RDT are then agreed with the end user.

4.11.2.6.3. What level of assembly of the product normally submitted to test?

The test is normally undertaken at the system level.

4.11.2.6.4. How many items are normally submitted to test?

As the test is done at the system level, generally one.

4.11.2.6.5. What type of product is appropriate for this test type?

Following the successful completion of the RGT program, the design authority may implement a reliability demonstration test (RDT) for complex and/or safety critical systems. The requirement for RDT for such systems is normally specified in the quality plan prior to the start of development. In such circumstances, it is necessary to initiate a robust RGT program to minimize the risk of "failure" during RDT.

4.11.2.6.6. What is the typical duration of a test program?

Typical test duration is months.

4.11.2.6.7. Should separate or combined environments be considered?

The RDT is based on a combined vibration/thermal stress environment, usually for a single stress cycle profile chosen to be typical of in-service. The actual stress level may be accelerated above "typical" in order to gain a corresponding reduction in test time. The eventual test profile and levels selected for the RDT are then agreed with the end user.

4.11.2.6.8. Are variations in product strength and environmental stresses considered?

Product strength and environment stress variability's are not considered.

4.11.2.6.9. Does the test supply knowledge on reliability parameters?

Yes, the test does supply knowledge on reliability parameters.

4.11.2.6.10. Does the test consume product life?

The test does consume some of the equipment's potential life.

4.11.2.6.11. What are the main norms, standards, technical references relevant to this test type?

MIL-HDBK-781, reliability design qualification and production acceptance tests. No longer maintained. Reliability Test Methods, Plans, and Environments for Engineering Development, Qualification, and Production, April 1996. This handbook contains test methods, test plans and environmental profile data presented in a manner which facilitates their use with tailorable tasks when appropriate. The testing of equipment procured for new military systems is an increasingly complex process. Test methods, test plans and test environments must be selected which will ensure that contractually required reliability levels are attained in the field and early defect failures are removed prior to field deployment. MIL-HDBK-781 provides a menu of test plans, test methods and environmental profiles. The most appropriate material may be selected for each program and incorporated into the tailored reliability test program. The handbook sections on reliability test methods and test plans present methods for growth monitoring, ESS, mean-time-between-failure assurance testing, sequential tests, fixed-duration tests and all-equipment tests, including a durability/economic life test. The sections on test environmental profiles provide typical test environments for fixed-ground equipment, mobile ground vehicle, shipboard, jet aircraft, turboprop and helicopter, and missiles and assembled external store equipment. The references provided will expand the user's knowledge and aid in the design and implementation of reliability test programs through more detailed data.

UK MOD, DEF-STAN – 00-42 (part 3/1): reliability and maintainability (R&M) assurance guidance, Part 3: R&M Case, 1999.

MIL-STD-785-Rev B Reliability Program For Systems And Equipment, September 1980, 88 pages. This military standard consists of basic application requirements, specific tailorable reliability program tasks and an appendix which includes an application matrix and guidance and rationale for task selection. Effective reliability programs must be tailored to fit program needs and constraints, including life cycle costs (LCC). This document is intentionally structured to discourage indiscriminate blanket applications. Tailoring is forced by requiring that specific tasks be selected and for those tasks identified, that certain essential information relative to implementation of the task be provided by the procuring activity. Many of the tasks solicit facts and recommendations from the contractors on the need for, and scope of, the work to be done rather than requiring that a specific task be done in a specific way. The selected tasks can be tailored to meet specific and peculiar program needs. Although not all encompassing, the guidance and rational provided in Appendix A is intended to serve as an aid in selecting and scoping the tasks and requirements.

4.11.2.7. *ROSE robustness testing*

4.11.2.7.1. What are the objectives of the test?

ROSE means robustness specification for environmental test. The test objective is to find weaknesses in an early design phase, which could start already in conceptual design and component specification phase. The key element is that components and products will be tested until they break in order to find a destruct level. With the knowledge of this level, estimations can be done for behavior and reliability of the design where the components are used or for the risks you face in designs when in use.

Companies often perform tests based on internal or external standards, like IEC. In that testing, fail criteria are based on pass/fail because the levels are stated and subject to test. By using the ROSE test philosophy, this approach is left and the product robustness is the subject.

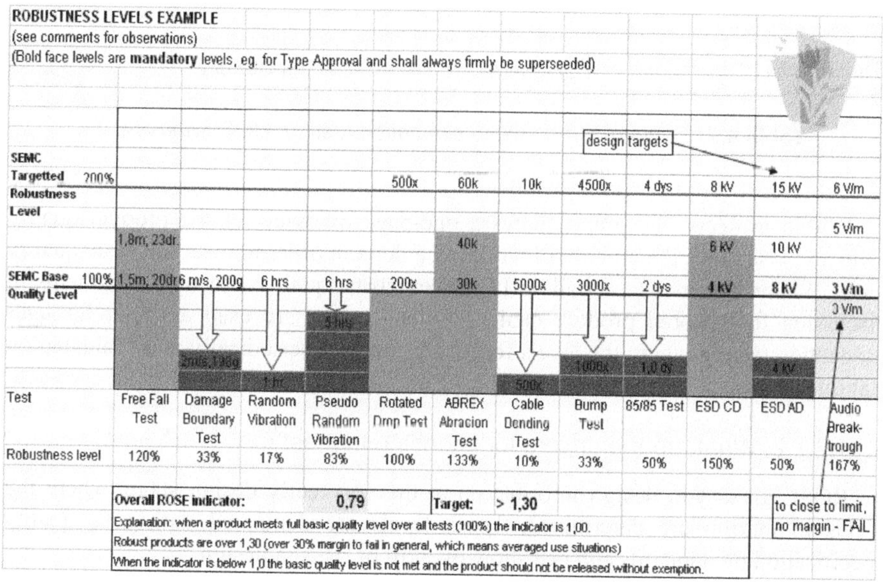

Figure 4.6. *Example of robustness levels. For a color version of this figure, see www.iste.co.uk/elhami/mechatronics.zip*

When knowing the fail level, a translation to the base quality level or standard is made easy and the margin between both could be an indicator of product reliability. This enables the quantification of reliability by means of this KPI (Key Performance Indicator).

ROSE is based on the step stress approach, where the stress is increased during testing. This is very similar to HALT and HA-ESS testing; however, conventional test methods, like the vibration test, drop test, temperature/humidity tests, are used. In addition, single stress is applied in order to find the correct root causes. This is the main difference with HALT, where six axis omnidirectional vibration is combined with rapid temperature changes and power cycling.

Simulation testing	ROSE testing	HALT testing
Longer testing, pass/fail, proven, standards Non-destructive RCA when necessary existing tests Single stress good simulation field experiences	Fast testing, levels, "new" Based on standards destructive (TTF) RCA always necessary existing tests Single stress	Fast testing, levels, "new" No standards destructive (TTF) RCA always necessary new test Combined stress
Three weeks	One week	One day

Table 4.2. *Comparison between simulation testing: ROSE and HALT*

Within projects, the graph shows a one-page overview of the product quality level and visualizes the amount of work to be done in order to meet the basic quality level. Where it comes to a trade-off between costs, quality and time, this helps in discussion for tailored product quality. Often, results are used as input for later production testing where the levels found are checked periodically and evaluated on stability.

4.11.2.7.2. What are the drivers for choosing the test?

ROSE gives you design information at the very early design stage, where the levels to margin are determined. These levels are compared with base quality specification or standards and can show overkill or risks in design.

4.11.2.7.3. What level of assembly of the product normally submitted to test?

ROSE is very helpful to rank components, boards and products.

4.11.2.7.4. How many items are normally submitted to test?

For a test done with ROSE philosophy, at least three samples should be used. This value is justified when using Weibull's distribution and when checking afterwards if the same fail mechanisms are found. If not, the test should be re-done with five or more samples in order to isolate the correct fail modes.

4.11.2.7.5. What type of product is appropriate for this test type?

Products that can be submitted are especially components, early designs/concepts and products. When applying ROSE in each stage, the reliability growth can be tracked. They can be electronic as well as mechanical products.

4.11.2.7.6. What is the typical duration of a test program?

The typical test duration is one to two days per test. For a full ROSE program, it will always be within one week.

4.11.2.7.7. Should separate or combined environments be considered?

In order to find and isolate root causes, ROSE testing needs to be done in separate environments. Combination of tests in a test program is possible, where sequencing is part of the consideration.

4.11.2.7.8. Are variations in product strength and environmental stresses considered?

Product strength is the topic of this test philosophy, especially in terms of robustness. Less in terms of resistability to the use environment, typical simulation tests are in place.

4.11.2.7.9. Why fault conditions or test severity are important

Because of single fault conditions and variations of the severity of test and test conditions, for example, pulse shape, it gives insight into the physics of failure and reliability parameters.

4.11.2.7.10. Does the test consume product life?

The tests are always destruction and consume full life potential of the product. However, because of the nature of this philosophy, it is not the simulation of product lifetime.

4.11.2.7.11. What are the main norms, standards, technical references relevant to this test type?

This type of testing is new, no standard is available yet. Presentations are available at www.ceees.org and www.plot.nl.

4.11.2.8. *MEOST robustness testing*

4.11.2.8.1. What are the objectives of the test?

MEOST means multiple environmental overstress testing and is described by Bhote and Bhote (2004). The intention is to address multiple environments in terms of mutual influences. This is different from sequential testing, where tests are done in a certain order.

MEOST can be seen as a successor to ROSE testing because the sequential testing can be addressed in ROSE, but not the mutual influences. MEOST could be closer to the real-world environment. However, it disables the tailored and direct root cause analysis.

Because of the overstress philosophy in MEOST, it still gives design information in terms of robustness. Ranking of components and products is a good opportunity, but slightly influenced by the way of testing and influences.

4.11.2.8.2. What are the drivers for choosing the test?

MEOST gives you design information, with a nod towards real-use environments. Still overstress testing is applied, so a matrix of stresses could help to find design weaknesses, but relation to standards is difficult.

4.11.2.8.3. What level of assembly of the product is normally submitted to test?

MEOST is very helpful to rank components, boards and products. For systems, it can be helpful to check these in a combined environment.

4.11.2.8.4. How many items are normally submitted to test?

For a test done with MEOST philosophy, at least five samples should be used. This value is justified when using Weibull's distribution and when checking afterward if the same fail mechanisms are found. It needs to be noted that sample sizes are higher because of a higher degree of freedom and variables in test.

4.11.2.8.5. What type of product is appropriate for this test type?

Products that can be submitted are especially products and systems.

4.11.2.8.6. What is the typical duration of a test program?

The typical test duration is one to two days per test. For a full MEOST program, it can be about two to three weeks.

4.11.2.8.7. Should separate or combined environments be considered?

MEOST is especially focused on combined environments and their mutual influences.

4.11.2.8.8. Are variations in product strength and environmental stresses considered?

Product strength is the topic of this test philosophy, especially in terms of robustness and resistability to the use environment. Because of the product/system test and multiple environments, there is a link to simulation testing.

4.11.2.8.9. Does the test supply knowledge on reliability parameters?

The basic reliability parameters are not addressed.

4.11.2.8.10. Does the test consume product life?

Yes, although the test might not be destructive, the levels are that high that products cannot be sold anymore and need to be amortized.

4.11.2.8.11. What are the main norms, standards, technical references relevant to this test type?

See Bhote and Bhote (2004).

4.11.2.9. *Conventional ESS*

4.11.2.9.1. What are the objectives of the test?

ESS means Environmental Stress Screening, where the keyword and main difference with ROSE and MEOST is in the word screening. This is not a test in order to find the product robustness level but a screen with the intention to check on product reliability stability. Therefore, it is often more manufacturing-oriented.

ESS is a product development process which involves the "precipitation and detection of hidden defects" by applying accelerated stresses of vibration, thermal cycling or power cycling. However, other definitions are known and often based on earlier test results.

ESS refers to the process of exposing a manufactured product or component (typically electronic) to stresses such as thermal cycling and vibration in order to precipitate latent defects in patent ones during the screening process. The surviving population, upon completion of screening, can be assumed to have a higher reliability than a similar unscreened population.

Developed to help electronic manufacturers detect product defects and production flaws, ESS is widely used in military and aerospace applications, less so for commercial products. The tests need not be elaborate, for example, switching an electronic or electrical system on and off a few times may be enough to catch some simple defects that would otherwise be encountered by the end user very soon after the product was first used. Tests may include the following:

- temperature variations;

- vibration tests;

- pressure;

- flexibility tests.

ESS can be performed as part of the manufacturing process or it can be used in new product qualification testing. An ESS system usually consists of a test chamber, controller, fixturing, interconnect and wiring, and a functional tester. These systems can be purchased from a variety of companies in the environmental test industry.

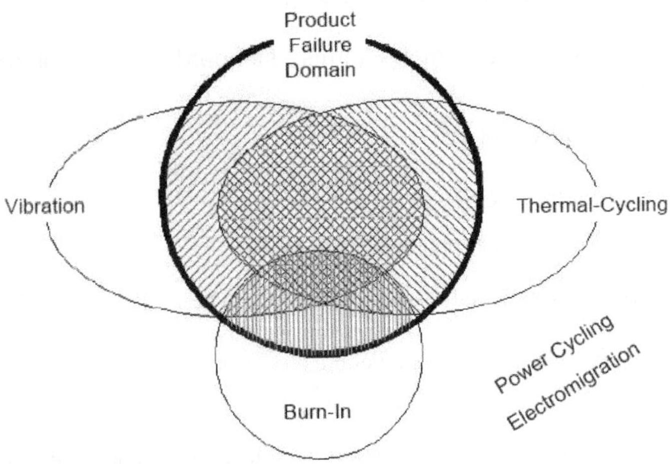

Figure 4.7. *Multi-level screening effectiveness. For a color version of this figure, see www.iste.co.uk/elhami/mechatronics.zip*

The stress screening from this process will help find infant mortality in the product. Finding these failures before the product reaches the customer yields better quality, lower warranty expenses and happier customers.

4.11.2.9.2. What are the drivers for choosing the test?

The drive is to eliminate the products presenting early defects bound to manufacturing and to the component dispersion. It prevents weak products, especially with infant mortalities, first part of bath tub curve, from entering the field.

4.11.2.9.3. What level of assembly of the product normally submitted to test?

ESS is applied to final products, mainly boards, equipment and systems. Components can be subjected to ESS, but often subjected to a burn in or a pre-conditioning program.

4.11.2.9.4. How many items are normally submitted to test?

Normally, the number of items tested is 100% of the production.

4.11.2.9.5. What type of product is appropriate for this test type?

All electronic products should be submitted to ESS.

4.11.2.9.6. What is the typical duration of a test program?

Normally, the test duration is in hours.

4.11.2.9.7. Should separate or combined environments be considered?

Both are popular. See the result of CEEES survey – CEEES Technical Advisory Board Reliability & ESS Publication (2009).

4.11.2.9.8. Are variations in product strength and environmental stresses considered?

Product strength and variabilities are not considered on the product level. Variations in production and incoming product quality are covered.

4.11.2.9.9. Does the test supply knowledge on reliability parameters?

The test does not produce knowledge on reliability parameters.

4.11.2.9.10. Does the test consume product life?

Very partially, and should be sufficiently low to be considered as acceptable.

4.11.2.9.11. What are the main norms, standards, technical references relevant to this test type?

There is a good guideline in the application of ESS in the NORTEL ESS guideline NT 011.

4.11.2.10. *Highly accelerated stress screening*

4.11.2.10.1. What are the objectives of the test?

The principle of the Highly Accelerated Environmental Stress Screening (HA-ESS) is to submit equipment coming out of production to levels of constraint far above the specified values, while staying below the destructive limits which could have been revealed by a campaign of highly accelerated life tests.

The most evident objective of screening is to discover and precipitate any weaknesses in the product; the consequences are to improve the operational reliability of the population of similar (same definition) items by precipitating latent failures and removing the weak items to improve the manufacturing processes to eliminate residual design defects.

Its main objectives are to:

– reveal more quickly the latent defects;

– detect more quickly the production process faults;

– identify the defects that may not have been revealed by a "conventional ESS" operation;

– reinforce the product maturity and robustness.

Other objectives can be associated; either for the supplier or for the customer:

– detect an unacceptable change in product quality;

– reduce the manufacturing and guarantee costs;

– lower the operational maintenance costs;

– obtain a better profitability during the guarantee period;

– establish or preserve the brand image of the manufacturer.

4.11.2.10.2. What are the drivers for choosing the test?

The drivers for choosing the test are time to market of robust product, safety consideration (avionics, automotive, etc.) and maintainability costs.

4.11.2.10.3. What level of assembly of the product is normally submitted to test?

Typically, sub-assemblies (normally not at the component level).

4.11.2.10.4. How many items are normally submitted to test?

The usual HA-ESS rule is to apply it to all the manufactured equipment. However, in the case of high volume production, such as the car industry, applying HA-ESS to 100% of the production is impossible for cadence as well as cost reasons. In that case, HA-ESS will be applied by sampling to detect the quality drifts (machines, components). It can be based on a dynamic sampling size.

4.11.2.10.5. What is the typical duration of a test program?

The duration of HA-ESS is product-dependent; however, because it is part of manufacturing process control, tests will not take longer than one day.

4.11.2.10.6. Should separate or combined environments be considered?

Combined environments are not a condition but can be employed if necessary to disclose a failure. For example, we can gradually increase the vibration level for a given rate of temperature variation or by gradually increasing this rate for a given level of vibration.

4.11.2.10.7. Are variations in product strength and environmental stresses considered?

Variations in product strength and environmental stresses are not considered.

4.11.2.10.8. Does the test supply knowledge on reliability parameters?

The test does not supply knowledge on reliability parameters.

4.11.2.10.9. Does the test consume product life?

Yes, the product under test is normally destroyed; however, the difference between the upper operating and upper destruct levels can be that high that for efficiency reasons the test is not performed destructively.

4.11.2.10.10. What are the main norms, standards, technical references relevant to this test type?

ASTE guide on HA-ESS "Environmental Stress Screening for Electronic Equipment using Highly accelerated Tests".

4.11.2.11. *Acceptance tests*

4.11.2.11.1. What are the objectives of the test?

Acceptance testing is black-box testing performed on a system (e.g. software, lots of manufactured mechanical parts, or batches of chemical products) prior to its delivery.

In some engineering sub-disciplines, it is known as functional testing, black-box testing, release acceptance, QA testing, application testing, confidence testing, final testing, validation testing, usability testing or factory acceptance testing.

In most environments, acceptance testing by the system provider is distinguished from acceptance testing by the customer (the user or client) prior to accepting transfer of ownership. In such environments, acceptance testing performed by the customer is known as beta testing, user acceptance testing (UAT), end-user testing, site (acceptance) testing or field (acceptance) testing.

4.11.2.11.2. What are the drivers for choosing the test?

A confluent application of a series of tests on an example of a product complying with a given definition specification file permits us to find the customer confidence in the quality of the product and is a green light to the transfer of ownership.

4.11.2.11.3. What level of assembly of the product is normally submitted to test?

All levels of assembly where a transfer of ownership exists.

4.11.2.11.4. How many items are normally submitted to test?

Few, often one exemplary per period of production.

4.11.2.11.5. What type of product is appropriate for this test type?

This test type is suitable for any product type.

4.11.2.11.6. What is the typical duration of a test program?

The overall test duration is usually measured in days.

4.11.2.11.7. Should separate or combined environments be considered?

Environmental conditions are normally applied separately.

4.11.2.11.8. Are variations in product strength and environmental stresses considered?

This test type does not consider variations in product strength.

4.11.2.11.9. Does the test supply knowledge on reliability parameters?

This test type does not supply knowledge of reliability parameters.

4.11.2.11.10. Does the test consume product life?

The intent with this test is not to consume product life.

4.11.2.11.11. What are the main norms, standards, technical references relevant to this test type?

No specific known norms.

4.11.2.12. *Production reliability acceptance test (PRAT)*

4.11.2.12.1. What are the objectives of the test?

The production reliability acceptance test (PRAT) is a process to demonstrate that the production standard hardware continues to meet the reliability requirements throughout production. This is normally achieved by subjecting a selected sample of production hardware to environmental conditions and functionally tested as specified in the PRAT plan that will be developed during pre-production.

Its purpose is to demonstrate successful subsystem operation in simulated field environments, and to verify compliance with contractual reliability requirements. This aggressive testing also confirms consistency in the product's manufacturing processes, and identifies changes in the component's quality or supplier's manufacturing processes.

Although performed at an accelerated rate, TAAF is a mirror image of the testing profiles used for PRAT, and its associated failure rates are based on the PRAT profiles observed. TAAF is conducted on equipment to promote reliability growth and identify areas for improvement within the current design. TAAF also identifies the operational limitations under which the equipment will work, which provides the company with a better assessment of design margins and a timely feedback of design, manufacturing process, test and/or supplier shortcomings.

4.11.2.12.2. What are the drivers for choosing the test?

Control the quality of production by accepting or rejecting a lot based on PRAT criteria.

4.11.2.12.3. What level of assembly of the product is normally submitted to test?

This process is applied at the system level assembly, containing all major units. The system will be subjected to a PRAT profile that simulates temperature excursions, vibration and application of power and functional tests at selected points in the cycle. The test profile is normally confirmed and proven during the development trials and agreed with the customer as part of the overall quality plan.

4.11.2.12.4. How many items are normally submitted to test?

The PRAT process involves testing the selected sample number of production “hardware” from each manufacturing production batch, dependent on the manufacturing throughput.

4.11.2.12.5. What type of product is appropriate for this test type?

In general, when there is a procurement policy requesting that type of test, that is, Government Furnish Equipment.

4.11.2.12.6. What is the typical duration of a test program?

The typical duration of PRAT testing is measured in weeks.

4.11.2.12.7. Should separate or combined environments be considered?

The test normally applies combined environments.

4.11.2.12.8. Are variations in product strength and environmental stresses considered?

PRAT testing does not address variations in product strength.

4.11.2.12.9. Does the test supply knowledge on reliability parameters?

PRAT does not supply knowledge of reliability parameters.

4.11.2.12.10. Does the test consume product life?

It consumes partially and can totally consume the life potential of the equipment.

4.11.2.12.11. What are the main norms, standards, technical references relevant to this test type?

MIL-STD-785-Rev B.

4.11.3. *Appendix 3: Feasibility test*

4.11.3.1. Purpose of a feasibility study

A feasibility study is designed to provide an overview of the primary issues related to a technical and business idea. The purpose is to identify any "make or break" issues that would prevent the test from being successful. In other words, a feasibility study determines whether the technical issues make sense, along with the business issues, but we are only concerned with the technical issues here.

A thorough feasibility analysis provides a lot of information necessary for the technical plan. For example, a good market analysis is necessary in order to determine whether the project is feasible. This information provides the basis for the market section of the business plan.

Because putting together a technical and business plan is a significant investment of time and money, you want to make sure that there are no major roadblocks facing you when the test plan is finalized and you make any investment in time and equipment. Identifying such roadblocks is the purpose of a feasibility study.

A feasibility study looks at three major areas:

– market issues;

– organizational/technical issues;

– financial issues.

Again, this is meant to be a "first-cut" look at these issues. For example, a feasibility study should not perform in-depth long-term financial projections, but it should perform a basic break-even analysis to see how much revenue would be necessary to meet operating expenses.

The purpose of the business plan is to minimize the risk associated with a new business and maximize the chances of success through research and maximize the chances for success through research and planning.

4.11.3.2. *Technological issues*

The cost and availability of technology may be of critical importance to the feasibility of a project, or it may not be an issue at all.

For example, a service organization, such as a childcare center, will have a few equipment issues and other technology-related issues to address. A manufacturing enterprise, on the other hand, may have a number of complex technology questions to analyze in order to determine whether or not the business is feasible.

Key questions to answer include:

– What are the technology needs for the proposed business?

– What other equipment does your proposed business need?

– Where will you obtain this technology and equipment?

– When can you get the necessary equipment?

– How does your ability to obtain this technology and equipment affect your start-up timeline?

– How much will the equipment and technology cost?

However, making wise decisions on even simple purchases such as office machines may require some research. Obviously, there are numerous types of personal computers on the market. You may want to check consumer reports for their recommendations, do some comparative shopping and ask acquaintances about their experiences with different companies. Your cost estimates will get plugged into your financial projections.

Naturally, the more complex the technology you need, the more research that will be required to make good decisions about it. This is important and should not be looked at trivially.

4.11.3.3. *The technical issues to look at during the feasibility study*

In the commercial market, we have to decide what we trying to achieve with a certain test. The main thrust is always the reliability of the product being manufactured.

The product life cycle begins at the component level and continues through the assembly level; the life cycle includes exposure to the following environments:

– assembly/process;

– testing;

– storage;

– transportation;

– operating;

– servicing (preventive and corrective maintenance).

4.11.3.4. Assembly/process

During processing and assembly of electronic assemblies, temperature excursions take place, for example, soldering including reflow, cleaning or imposed thermal cycling tests. These temperature excursions can be damaging to some parts of the assembly and consume some part of the available life. Keeping the number of excursions to a minimum should minimize this damage and the damage needs to be considered in the overall reliability estimates.

4.11.3.5. *Testing*

When devising the overall reliability test plan, the frequency of the applied test to the device and the fatigue that the testing induces into the product will be identified.

4.11.3.6. *Storage*

Storage tests will be defined by the product specification. The storage life of the product will be identified which may range from a few months to many years. A storage life cycle may include storage in a mobile unit to storage in a temperature-controlled warehouse.

4.11.3.7. *Transportation*

Transportation is not a single event that is simply defined. Each vehicle type may impose unique environmental loads. Each may also provide protection from certain environmental conditions. The duration and frequency of occurrence will influence how environmental effects are accumulated over time. The transportation configuration and degree of loading can also affect the loads imposed on the product.

4.11.3.8. *Operating*

It is assumed that the operational characteristics will not affect the reliability test, as the unit is a low power device and operational tests.

4.11.3.9. *Environmental Stress Screening (ESS)*

ESS has the potential to identify latent defects that may cause early failures in a product. ESS needs to be specifically designed to accelerate the failure of "weak" elements in the assembly. ESS does not add to the number of such failures but causes them to occur in a significantly shorter period of time.

Effective ESS programs should be supported by a well-planned "root cause analysis" (RCA) and corrective action resources. These enable timely corrective actions, elimination of latent defects and ultimate removal of the ESS process subject to continued monitoring and control of the manufacturing process.

4.11.3.10. *Conclusion*

All the above tests will be scrutinized during a feasibility study, equipment, infrastructure, human resources and all the associated costs that go with these requirements. This is the first part in looking at a reliability test program.

4.11.4. *Appendix 4: Comparison of ESS and HA-ESS*

This section evaluates the efficacy of the conventional ESS and of the highly accelerated ESS.

4.11.4.1. *Conventional Environmental Stress Screening method*

Environmental stress screening, as defined in the ASTE document of 1986, is referred to as conventional environmental stress screening as opposed to the "highly accelerated environmental stress screening" (HA-ESS). Experience shows that, for a serially produced item, some items are much more fragile than others. Essentially, there are two statistical distributions, the one relative to good products, the other to fragile products. The healthy population is located around a mean value situated beyond the specification.

In the case of conventional ESS use, highly accelerated tests are not necessarily conducted during the product development phase and the margin between the population and the specification is not known. In this method, the constraints are generally applied separately, and the laws of defect appearance are known in principle. It will also be noted that the constraint level is always lower than or equal to the specification value.

It is recognized in electronics that every product group will contain latent defects which will age faster than every healthy set (fuse effect). After conventional ESS testing, the healthy population has a slightly decreased life. On the other hand, the products with latent defects are mostly removed because the majority fail during a test. The effects on an untested population shown in Figure 4.8 due to ESS are illustrated in Figure 4.9. However, as the constraints are relatively low and the technical margins unknown, the danger is that some products with latent defects (weaknesses) could still be identified as being good.

4.11.4.2. *HA-ESS method*

In this method, as previously mentioned, it is essential to have previously performed a campaign of highly accelerated tests and to have established a robust design and a robust manufacture.

It is certain if not easy to demonstrate that the consequences of this highly accelerated ESS on the product will be more important than the one generated by conventional ESS. The test will reduce the mean of the good product distribution towards the specification to a greater extent than conventional ESS. It will of course be necessary to verify that the consequential life reduction is not important.

For the products with latent defects, the high levels applied increase the proportion of the failures revealed to a greater extent than with a conventional ESS.

The effects on the population due to HA-ESS are illustrated in Figure 4.3. This figure shows a greater shift of the population mean for the latent (abnormal) failure distribution than that of the normal population. The latent failure distribution is rejected, as not being compliant with the specification. As the mean shift is greater than with conventional ESS, a greater number of weaknesses are revealed.

4.11.4.3. *Comparison of the ESS methods*

Table 4.3 summarizes the differences between the main characteristics of the conventional ESS and of the highly accelerated ESS.

	Conventional ESS	Highly accelerated ESS
Objective	Eliminate the products presenting early defects bound to manufacturing and to the component dispersion	
Principle	The application of constraints within the limits of the product specification	The application of constraints beyond the specification. Note: allows us to detect the components' dispersion
Input data	The product specifications and its life profile	Knowledge of the operation margins and of the product limitations. Note: the product robustness has been improved by increasing the operation margins (tolerance increased in component dispersal)
Responsibility	Manufacturing only	R&D and manufacturing

Constraints application mode	Sequential type: temperature, vibration	Combined: temperature, mechanical shocks, electric constraints, etc.
Efficiency	Moderate or weak on the stabilized products	Good: increased efficiency in terms of control of the operation margins Maturation and faster mastery of the manufacturing process
Profile acquisition	Long and iterative because based on the return of experience – 1 year or more	Fast due to the methodology – approximately 3–6 months
Speed	Indicative duration: 2–4 days	Indicative duration: 2–8 hours (reduces the production cycles)

Table 4.3. *Differences between the main characteristics of conventional ESS and highly accelerated ESS*

4.11.4.4. *Other advantages of HA-ESS related to ESS*

In addition to the benefits already mentioned previously, the implementation of the highly accelerated ESS presents the following advantages:

– a standard process applicable by the company both internally and externally;

– the facilitation of the dialogue between the design actors and the manufacturing factors;

– the satisfaction for the employees to be better involved in the possibility to touch the state of the art of the manufacturing process;

– a systematic approach of quality.

4.11.4.5. Potential problems and precautions

To set up a highly accelerated ESS operation within the manufacturing phase can be delicate, especially at the level of the communication within the company community. It therefore requires a certain number of pre-requisites. These include the prior implementation of highly accelerated tests in order to establish the highly accelerated ESS profile. Within this, it is necessary to involve the design teams early in the process. It is also necessary to increase the subcontractor's awareness in the method, and to establish with them appropriate procedures.

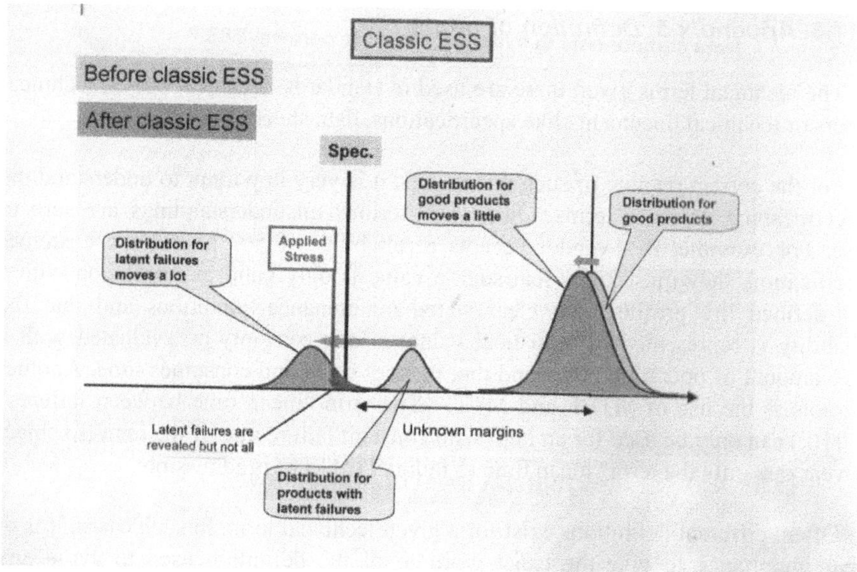

Figure 4.8. *Effects on the product population of conventional ESS testing. For a color version of this figure, see www.iste.co.uk/elhami/mechatronics.zip*

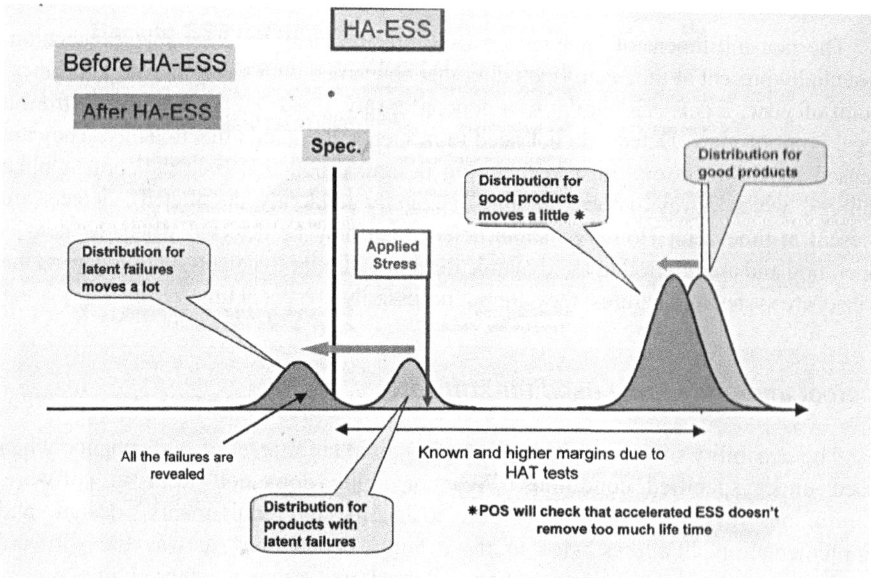

Figure 4.9. *Effects on the product population of conventional HA-ESS testing. For a color version of this figure, see www.iste.co.uk/elhami/mechatronics.zip*

4.11.5. *Appendix 5: Definition of terms*

The technical terms given there are used in standards, technical books, technical papers or technical documents like specifications, data sheets, etc.

For the correct reading of such documents, it is very important to understand the exact meaning of those terms. Otherwise, serious misunderstandings are sure to arise. For example, if a vendor accepts a reliability of 500 hours in a customer specification, they must know that such a value is only valid in correlation with a well-defined life profile, with clear stated maintenance conditions and that the reliability is represented by statistical values which can only be evaluated with a huge amount of operating hours and that induces costs and consumes time. Another example is the use of MTBF and MTTF. The term "mean time between failures" (MTBF) can only be used for an item with constant failure rate. If the item is subject to wear out, only the term "mean time to failure" (MTTF) is admissible.

Often, different definitions exist for a given technical term. In such cases, it is of great importance to give the exact wording of the definition used to avoid any misunderstanding.

Defect

The non-fulfillment of an intended usage requirement or reasonable expectation, essentially present at time zero (including that concerned with safety). From a technical point of view, a defect is similar to a nonconformity, however, not necessarily from a legal point of view. Defects do not need to influence the item's functionality. They are caused by flaws (errors, mistakes) during design, production or installation. Unlike failures, which always appear in time (generally randomly distributed), defects are present at time zero. However, some defects can only be detected when the item is operating and are referred to as "dynamic defects". Similar to defects, with regard to the cause, are systematic failures; they are not necessarily present at time zero.

Defect immunity (only used for software)

The capability of the software product to maintain a level of performance when used under specified conditions. Wear or aging does not occur in software. Limitations in "reliability" are due to faults in requirements, design and implementation. "Failures" due to these faults depend on the way the software product is used and the program options selected, rather than on elapsed time.

Dependability

A collective term used to describe the availability performance and its influencing factors, such as reliability performance, maintainability performance and logistical support performance. Dependability is used only for general descriptions in non-qualitative terms. It is one of the time-related aspects of quality. The definition of dependability given above is taken from IEC 50(191), which also includes related terms and definitions.

Durability

The capacity of an item to execute a required function in specified conditions of use and of maintenance until the end of useful life or until it no longer meets with its expected economical or technological requirements

Environmental stress screening (ESS)

A screening process, based on a stimulation or a set of stimulations intended to remove defective items, or those likely to exhibit early failures. The applied stress levels are normally similar to those expected in field operation. ESS is often performed at the assembly level (PCB) or equipment level on a 100% basis to find defects and/or systematic failures during a preproduction phase or to provoke early failures during the series production and therefore to improve the reliability of a lot of items by elimination of the defectives one. To be most effective, ESS should be tailored to the kind of item and the production process.

Failure

The termination of the ability of an item to perform a required function.

Failure identification, analysis and fix

A general concept where the cause of failure is isolated, analyzed and fixed. A number of testing procedures can be used in this method. These procedures can be applied in parallel with the existing development tests or as dedicated TAAF (test, analyze and fix) or TAAR (test, analyze and redesign) procedures. These are closed-loop reliability growth methodologies. The purpose is not to prove that a reliability goal has been met, but rather to deliberately search out and eliminate deficiencies. In TAAF/TAAR, failures are welcome. The TAAF/TAAR concept is necessary because, even with the very best of modern engineering methods and design tools (CAD, CAE), initial designs for mechanical or electronic systems that are complex or that involve new technology have reliability deficiencies that are difficult to fully detect and eliminate through design analysis. To summarize, the maturation of revolutionary products (which includes many new items in designs, technologies and conditions of use) requires experimentation.

Failure rate λ(t)

The failure rate of an item is the probability (referred to as dt) of a failure in the interval (t, t +dt) under the condition that the item was new at t = 0 and did not fail in the interval (0,t).

Fault

The state of an item, characterized by the inability to perform a required function, excluding inabilities due to preventive maintenance, other planned actions or lack of external resources. A fault can be a defect or a failure, having thus as possible cause a flow (error, mistakes) during design, production or installation (for defect or systematic failure) or a failure mechanism (for failures).

Fault mode, effect and criticality analysis (FMECA)

A qualitative/quantitative method of reliability analysis which includes the analysis of fault modes and effects (FMEA) while considering for each fault mode the probability of occurrence and ranking of its severity. The goal of an FMEA or FMECA is to determine all potential hazards and to analyze the possibilities of reducing their effects or their probability of occurrence. All possible failure and defect modes and causes have to be considered from the lowest to the highest integration level of the item considered (FMECA was formerly used for "Failure Mode, Effect and criticality Analysis" by considering only the possible failure modes and corresponding effect and criticality).

Fault modes and effect analysis (FMEA)

A quantitative method of reliability analysis, which involves for each element of an item the investigation of possible fault modes, and of the corresponding effects on other elements as well as on the required function(s) of the item (see also FMECA).

Fault tree analysis

An analysis to determine which fault modes of the elements of an item and/or which external event may result in a stated fault mode of the item considered, presented in the form of a fault tree.

Functionality (capability, technical performance)

The capability of a product to provide the required function when it is used under specified conditions. Alternatively, the ability of an item to meet a service demand of the stated quantitative characteristics under the given conditions.

Life cycle

This consists of all phases of the life of an item including acquisition, operation, storage, maintenance and disposal.

Life cycle profile

The sum of all states of an item under the given conditions over the complete life cycle of the item (the life cycle profile defines the required functions and the environmental conditions during the different life cycle phases as a function of time).

Life extension analysis

This evaluates the ability of a material to meet extended life, role and deployment changes. The process is presented in this AECTP, which is a framework document. The purpose of this document is to acquaint project (program) managers with the engineering principles involved when evaluating the implications of extended life requirements, and also with the outline of a management tool that systematically addresses the issues to be resolved. Additional leaflets are included to provide further guidance for the engineering practitioner when addressing the detailed technical issues.

Lifetime

The time span between the start of operating and a failure of a non-repairable item. For repairable items, the term "useful life" is used.

Logistical support

All activities undertaken to provide effective and economical use of a product during its operating phase.

Maintainability

The probability that preventive maintenance or repair of an item will be performed within a stated time interval for given procedures and resources. A qualitative definition is the ability of a product to be retained in or restored to the ability to perform its required function in a given time interval under stated procedures and resources. Maintainability is subdivided in serviceability (preventive maintenance) and reparability (corrective maintenance or repair). The relationship between maintainability and other reliability measures is shown in Figure 4.10.

Mean time between failures (MTBF)

MTBF = $1/\lambda$ is the expected value (mean) of the item's failure-free operating time under the condition that the item was new by starting the operating time. MTBF should only be used for items with constant failure rate λ. For items with a time-dependent failure rate $\lambda(t)$, the term MTTF (mean time to failure) will be used.

Mean time to failure (MTTF)

The expected value (mean) of an item's failure-free operating time under the condition that the item was new by starting the operating time.

Mean time to mission failures (MTTFM)

The expected value (mean) of the operating time of an item without any failure which would prevent the fulfillment of a defined mission (failures not having an influence on the mission are not relevant for MTTFM).

Mean time to preventive maintenance

The expected value (mean) of an item's preventive maintenance time.

Mean time to repair

The expected value (mean) of an item's repair time.

Mission profile (MP)

A specific task that must be fulfilled by an item during a stated time under the given conditions (the mission profile defines the required functions and the environmental conditions as a function of time).

Operational availability, point availability

The probability that a product will perform its required function under the given conditions at a stated instant of time.

Operational effectiveness

The capability of an item to fulfill the specified functional requirements.

Packaging, handling, storage and transportation

These activities include resources and procedures to ensure that all equipment and support items are preserved, packaged, packed, marked, handled, transported and stored properly for short- and long-term requirements. It includes material-handling equipment and packaging, handling and storage requirements, and pre-positioning of material and parts. It also includes preservation and packaging level requirements and storage requirements (e.g. sensitive, proprietary and controlled items). These activities include planning and programming the details associated with the movement of the system/product in its shipping configuration to the ultimate destination via transportation modes and networks available. It further includes the determination of critical engineering design parameters and constraints that must be considered during system development (e.g. width, length, height, component and system rating, and weight).

Qualification

The process of demonstrating whether an entity is capable of fulfilling specified requirements.

Reliability

The probability that a product will perform is a required function under the given conditions for a stated time interval. This does not mean that redundant parts may not fail. Such parts can fail and be repaired. The relationship between reliability and other reliability measures is shown in Figure 4.10.

Safety

The ability of a product to cause neither injury to persons nor significant material damage or other unacceptable consequences. Alternatively, a state in which the risk of harm (to persons) or damage is limited to an acceptable level. The relationship between safety and reliability measures is shown in Figure 4.10. Safety is subdivided into accident prevention (the product is safely working while it is operating correctly) and technical safety (the product has to remain safe even if a failure occurs). Safety is one of the aspects of quality. The alternative definition is valid for the purposes of quality standards. The term “safety” is defined differently in ISO/IEC guide 2.

Security

The capability of the software product to protect information and data so that unauthorized person or systems cannot read or modify them and authorized persons or systems are not denied access to them.

Supportability, maintainability (for software)

The capability of the software product to be modified. Modifications may include corrections, improvements or adaptations of the software to a change in environment, and requirements and functional specifications.

System effectiveness

The characteristic of an item to fulfill the required functions with the best possible ratio between reliability and life cycle costs.

Time between preventive maintenance (TBPM)

A predetermined time interval between two preventive maintenance actions. An item may require several different TBPM. In this case, the shortest one is specified and normally the longer TBPMs are multiples of the shortest one.

Usability

The capability of a product to be understood, learned, used and efficient for the user, when used under specified conditions.

Useful life

The total operating time of a product, ending for a non-repairable product when the failure probability becomes too high or the product functionality is obsolete, and for a repairable product when the intensity of failures becomes unacceptable or when after a failure the product is considered to be no longer repairable. The term "lifetime" is only used for non-repairable products. It is the time span between the initial operation and failure of a non-repairable item.

Validation

The process of confirmation by examination and provision of objective evidence that the particular requirements for a specific intended use are fulfilled. In design and development, validation concerns the process of examining a product to determine conformity with the manufacturer's specifications. Validation is normally performed on the final product under defined operating conditions. It may be necessary in earlier stages. The term "validated" is used to designate the corresponding status. Multiple validations may be carried out if there are different intended uses. Validation is done against the "product specification".

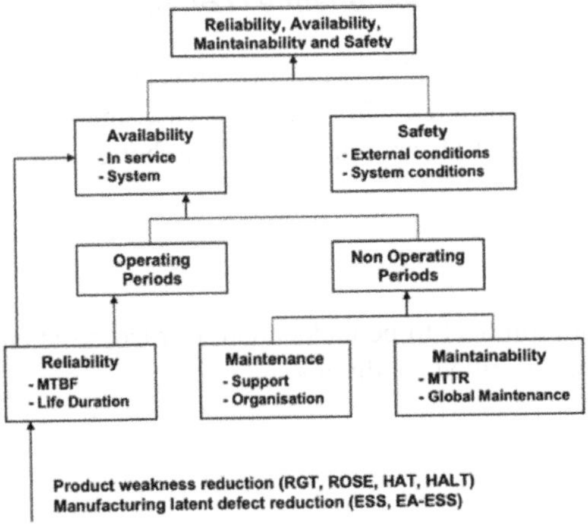

Figure 4.10. *Hierarchy of terms*

4.11.6. *Appendix 6: About MTBF calculations based on HALT results*

If we refer to the title of the paper "From HALT results to an accurate field MTBF estimate" (McLean 2010), we may think that it is possible to estimate the product field failure rate from the results of HALT data.

Does the nature of the HALT approach really allow this?

The probability density function of a Weibull random variable x is:

$$f(x;\lambda,k) = \begin{cases} \frac{k}{\lambda}\left(\frac{x}{\lambda}\right)^{k-1} e^{-(x/\lambda)^k} & x \geq 0, \\ 0 & x < 0, \end{cases}$$

where k > 0 is the shape parameter and λ > 0 is the scale parameter of the distribution. Its complementary cumulative distribution function is a stretched exponential function. The Weibull distribution is related to a number of other probability distributions; in particular, it interpolates between the exponential distribution (k = 1) and the Rayleigh distribution (k = 2).

The hidden assumption in the article by Harry McLean titled "From HALT Results to an Accurate Field MTBF Estimate" is that the shape parameter is constant whatever the stress level applied to the unit under test.

Demonstration of this assumption supposes long duration tests which is the contrary of the seeking objective (i.e. shorten the test duration).

This assumption can be verified in particular for a failure mode following the exponential distribution where k = 1; some authors consider that for a complex system constituted of many different sub-assemblies, the trend of the system-associated distribution is exponential one. But it could not be generalized. Otherwise, the accelerated tests would have been replaced by highly accelerated tests… which is not the case.

In other words: with accelerated tests, we are in the known world (in a limited domain of stress amplitudes: the model of acceleration is known, the associated parameters are known, the invariance of the shape parameter is supposed); in the case of aggravated tests (so called highly accelerated tests), we are in an unknown world, but still assume that the correction of the identified defects, even in the case they are not relevant to increase the margins, will not degrade the product.

And about when is a product mature? => This is challenging – not necessarily clear even at end-of-life… Problems may rise due to problems in hw–sw joint operation. Suppliers or subcontractors may make changes without any announcement and infant mortality issues may be repeated.

This is more part of product quality assurance/control. Integration of HALT in production lines (HASS), requalifications or market reviews, as well as incoming inspections, can reduce the risk of infant mortalities.

A mature product is a product characterized by a completeness of industrialization (production line operational, testing tools available and industrialization having achieved a good level of FPY (first pass yield) and having disclosed in operational conditions a reliability close to its inherent reliability). The inherent reliability is a measure of the overall "robustness" of a system or piece of equipment. It provides an upper limit to the reliability and availability that can be achieved. In other words, no matter how much inspection or maintenance you perform, you will never exceed the inherent reliability. It corresponds to the current state of the art.

4.12. References

Birolini, A. (1997). *Quality and Reliability of Technical Systems*. Springer-Verlag, Berlin, Heidelberg.

BNAE RG Aéro (1991). RG AERO 40 – Recommandation générale pour la Spécification Management de Programme – 6/91, edition : IMDR, ref. 68 [Online]. Available at: https://www.imdr.eu/shop_818-46673-5302-836/rg-aero-40-recommandation-generale-pour-la-specification-management-de-programme-6-91.html.

Bhote, K. and Bhote, A. (2004). *World Class Reliability: Using Multiple Environment Overstress Tests to Make it Happen*. AMACOM.

CEEES (1996). Synthesis of an ESS survey at the European Level [Online]. Available at: http://www.ceees.org/downloads/papers/pdf/CEEES_Publication_No3.pdf.

CEEES (2003). List of dedicated or related documents to ESS [Online]. Available at: http://www.ceees.org/downloads/papers/pdf/CEEES_Publication_No4.pdf.

CEEES (2009). Reliability – For a mature product from the beginning of useful life. The different type of tests and their impact on product reliability. CEEES Technical Advisory Board Reliability & ESS Publication, 9.

CIN EG 1 and French Ministry of Defense (1999). Guide pour la prise en compte de l'environnement dans un programme d'armement [Online]. Available at: https://aste.asso.fr/wp-content/uploads/cineg01.pdf.

Fragola, J.R. (1996). Reliability and risk analysis data base development: An historical perspective. *Reliability Engineering & System Safety*, 51(2), 125–136.

ISO (2000). ISO9000:2000. Quality management systems – Fundamentals and vocabulary edition 2. December, ISO/TC176/SC1.

ISO/IEC (2001). ISO/IEC 9126-1, Software Engineering – Product Quality – Part 1. ISO/IEC JTC 1/SC 7.

Kahramanoğlu, G. (2010). Field return rate estimation process [Online]. Available at: http://www.ieee-astr.org/kahramanoglu_ASTR2010.pdf.

McLean, H. (2009). *HALT, HASS, and HASA EXPLAINED Accelerated Reliability Techniques*, Revised edition. Asq, Milwaukee, WI.

McLean, H. (2010). From HALT results to an accurate field MTBF estimate. *2010 Proceedings – Annual Reliability and Maintainability Symposium (RAMS)*. IEEE.

Toshiba (n.d. a). Testing [Online]. Available at: http://www.semicon.toshiba.co.jp/eng/product/reliability/device/testing.

Toshiba (n.d. b). Estimation [Online]. Available at: http://www.semicon.toshiba.co.jp/eng/product/reliability/device/estimation.

USA Department of Defense (n.d.). Defense specifications. MIL-STD-961.

USA Department of Defense (1991). Reliability prediction of electronic equipment. Military handbook, MIL-HDBK 217F.

USA Department of Defense (2020). Defense and program-unique specifications. Military of Defense standard practice, MIL-STD-961E.

Wilkins, D.J. (2002). The bathtub curve and product failure behavior part one – The bathtub curve, infant mortality and burn-in. *Reliability HotWire*, 21 [Online]. Available at: http://www.weibull.com/hotwire/issue21/hottopics21.htm.

5

Reliability Climatic Test for Composites Based on a Probabilistic Arrhenius Model

This chapter puts forward a methodology to verify the reliability of composite products over their lifetime using a probabilistic Arrhenius model. It will present the challenges that the automotive industry faces and study the potential improvements to be considered in order to improve reliability during the design and validation phases. The formulas developed in this research allow us to set up an accelerated test for the strict necessary needs. The goal is the integration of variability from activation energy (E_a), test conditions (test bench temperature, T_b) and a dispersion from usage temperatures (T_u).

This chapter is structured into four sections:

– automotive reliability context (reliability: needs and constraints, Arrhenius model presentation);

– probabilistic Arrhenius model;

– concrete case application: automotive application;

– the conclusion analyses the limits and perspectives on this subject.

5.1. Introduction: needs and constraints of automotive reliability

Reliability and the automotive world are intimately linked by their histories (Delaux et al. 2017a, 2017b), their current challenges and their futures. This section will highlight these three aspects taking into account the global context of this ever-changing industry.

Chapter written by David DELAUX, Thomas ILLING and Abdelkhalak EL HAMI.

There are many publications on reliability with each one casting a different light on the theme (Bertsche 2008; Verma et al. 2010; Delaux et al. 2017a, 2017b). Nonetheless, the automotive world is in a permanent search for innovation, production, mobility, autonomy and speed. In 1900, the United States had only 8,000 cars and a few hundred kilometers of road. Currently, there are more than one and a half billion motor vehicles and millions of kilometers of good roads worldwide. This economy uses and develops techniques and methods, increasingly important to the assurance of vehicle owners, in terms of total safety and comfort.

The methods of reliability testing, which estimate the probability of a system to survive under certain conditions, have in particular drawn upon the statistical theories of the 18th, 19th and 20th centuries. There are probabilistic approaches according to the stochastic diffusion (Prabhu 1965; the Normal Law) in 1777 by Pierre-Simon Laplace and in 1809 by Carl Friedrich Gauss, the approach of an exponential distribution in explicit forms, among others, by Leonhard Euler in 1748 in his book *Introductio in analysin infinitorum* or the proportional effect extolled by R. Gibrat in 1930 (Gibrat 1930; Aitchison and Brown 1957; the lognormal law).

The probabilistic formulations of the estimate reappear in the movement of Karl Pearson's school with Ronald Aymler Fisher (1890–1962) and William Sealy Gosset – alias Student – (1876–1937), who were all confronted with statistical studies based on insufficient data sample size. Ronald Fisher is generally presented as the father of Estimation Theory according to the principle of "maximum likelihood" (introduced in 1912 and developed until 1922–1925; Yadolah 1993; Damien Rodat et al. 2018).

Much later in 1939 and then again in 1951, professor and engineer Waloddi Weibull (Weibull 1951; Nosratabadi et al. 2017) published a statistical formulation that would surely become the most popular probabilistic distribution estimation for the processing of survival data: the Weibull distribution. It is not surprising to note that this statistical tool is still widely used in the automotive world, given the fact that Professor Weibull worked for major industrials like SAAB, the car manufacturer.

It was noted that the purpose of Professor Weibull's work was not to develop a theoretical statistical model or basic research, he sought to solve real physical problems, such as those related to the resistance of materials, fatigue and breaking points (Lannoy 2008). Here, we find the very nature of reliability, namely a multiple overlap of various sciences such as chemistry, physics, electrics, electronics, fatigue, thermo-mechanics, electromagnetism and pyrotechnics.

In the automotive world, the evolution of methods and statistical tools dedicated to reliability, as in other industries, was at the same time followed in parallel by an evolution of test methods to demonstrate the performance of the experimental reliability of the vehicle. Figure 5.1 illustrates the various and extreme applications of the first Model T Ford in 1915 and 1926.

Figure 5.1. *Model T Ford in a forest in 1915 and on the snow in 1926 (Wikimedia Commons 2022, 2020a)*

The complexity of the modern car adds new reliability challenges.

The Mercedes concept car shown in Figure 5.2, known by its codename F015, unveiled in 2015 in Los Angeles, is a fine example of ultra "electronized" and computerized high technology. This concept would allow for an all-electric propulsion system with a range of 1,000 km and driven autonomously.

Figure 5.2. *F015 Mercedes concept car from 2015 (Wikimedia Commons 2020b). For a color version of this figure, see www.iste.co.uk/elhami/mechatronics.zip*

The vehicles currently available (2014–2018) offer a truly impressive system of complexity. For example, most current models have no less than 100 electronic control units (ECUs), each with a computing power of more than 20 computers. The ECUs manage the entire vehicle, from comfort (e.g. air conditioning, lighting) to the management of the engine (e.g. fuel consumption, pollution) as well as certain safety features (e.g. ESP, airbags).

The number of ECUs in modern cars is equivalent to the number found in the first airbus (Charette 2009)!

According to a study by Toyota (Toyota, n.d.), it is estimated that there are more than 30,000 components in a standard vehicle today.

As stated by Automakers Turn to Plastics for Lighter Vehicles, "modern-day plastics account for roughly 50 percent of the volume of a typical vehicle, but only 10 percent of its weight" (Plastic Makes it Possible 2015). Composites, especially short glass fiber-reinforced polymers like Polyamide 6 with 30% glass fibers (PA6 GF30), are relevant materials that are increasingly being used in our vehicles. Figure 5.3 illustrates several aging processes that impact this kind of material.

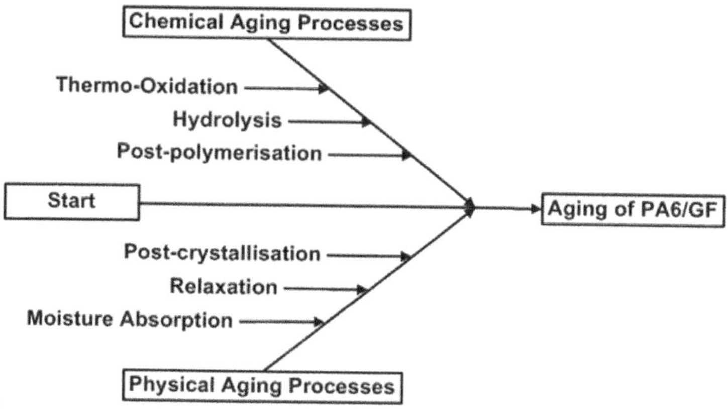

Figure 5.3. *PA6: relevant aging processes (Illing et al. 2017)*

One way to assess the reliability performance of complex systems in composite material is the environmental climatic tests. To accelerate this environment, reliability engineers use the Arrhenius model.

5.2. Proposition of a new probabilistic Arrhenius model

5.2.1. *Constant Arrhenius model*

Svante Arrhenius, recipient of the Nobel Prize in Chemistry in 1903, suggested a new formulation in his article called "On the velocity of the inversion of cane sugar by acids" in 1889 (Arrhenius 1889). This formulation gave the famous Arrhenius equation.

The Arrhenius relationship models the effect of temperature on the rate of a first-order chemical reaction and can be written as (Yang 2007):

$$\mathrm{v} = A_0 \exp\left(-\frac{E_a}{\mathrm{kT}}\right) \qquad [5.1]$$

where v is the chemical reaction rate in moles per second, E_a is the effective activation energy in electron-volts (eV), k is the Boltzmann constant (k = 8.61733 x 10–5 eV/◦C), T is the absolute temperature (the Celsius temperature plus 273.15 degrees), and A_0 is a constant related to material characteristics.

This speed expresses the amount of reaction per unit time (Demir and Cronin 2004). We can consider that a failure occurs when a critical amount (in moles) of reactant is reached. Therefore, the time to reach the critical amount is called "time to failure". Since the time to failure is proportional to the reciprocal of the reaction rate, we have:

$$\mathrm{L} = \mathrm{A}\exp\left(\frac{E_a}{\mathrm{kT}}\right) \quad [5.2]$$

where L is the life and A is a constant that depends on material properties, failure criteria, product design and other factors. Equation [5.2] is called the Arrhenius life relationship.

Activation energy is an important concept associated with the Arrhenius relationship. It is understood that a chemical reaction is the result of the collisions between the reactant molecules. The collisions take place very frequently, but only a small fraction of the collisions convert reactants into products of the reaction. The necessary condition for the collision to cause a reaction is that the molecules carry a minimum amount of energy to break bonds and form products. The minimum amount of energy is called the activation energy; it poses a barrier for molecules to climb over. The higher the activation energy, the lower the reaction rate and the longer the life. The activation energy is unique to a reaction, which determines the failure mechanism leading to failure. Therefore, each failure mechanism usually has different activation energy even for the same component. For example: the polyamide PA6 or PA66 are sensitive to thermal oxidation, post-crystallization and absorption with a unique E_a between 0.73 eV and 0.97 eV for each type of reaction (Eriksson 1998; Illing et al. 2017).

For composite materials like polymers, the effective activation energy is expressed in kJ/mol. In this chapter, eV is used as the base.

The conversion is defined by:

$$1\, eV = 96.4869\, kJ/mol$$

The graph shown in Figure 5.4 represents the E_a dispersion for different composite materials obtained with 56 E_a values compiled in 11 publications such as Gillen et al. (2005a, 2006) and Bernstein (2010).

The median of this lognormal distribution shows a probability of 0.85 eV with a coefficient of variation of 42% (statistical indicator based on ratio of mean and standard deviation).

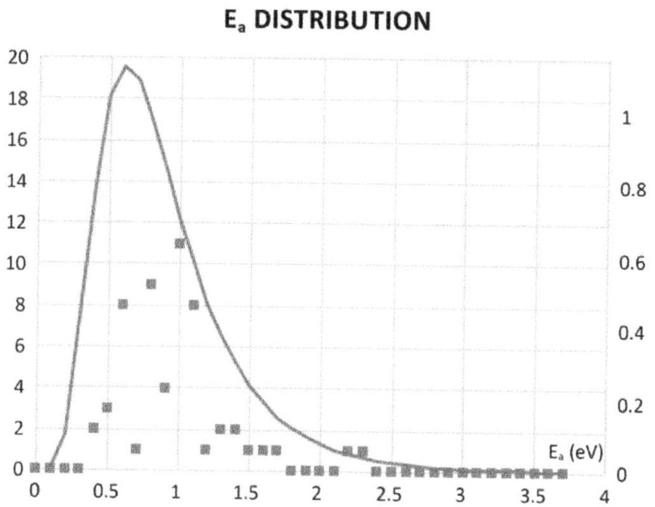

Figure 5.4. *E_a (eV) dispersion. For a color version of this figure, see www.iste.co.uk/elhami/mechatronics.zip*

The coefficient of variation is defined as (Koopmans et al. 1964):

$$CV_{ln} = \sqrt{e^{\sigma_{ln}^2} - 1} \qquad [5.3]$$

σ_{ln} is the standard deviation of lognormal distribution.

Since the Arrhenius life can represent a percentile, the acceleration factor A_f between the life L at temperature T and the life L' at temperature T' is as follows:

$$A_f = \frac{L}{L'} = \exp\left[\frac{E_a}{k}\left(\frac{1}{T} - \frac{1}{T'}\right)\right] \qquad [5.4]$$

The acceleration factor increases with the activation energy. If E_a is equal to 0.5 eV, as the temperature increases every 10°C, the life would reduce approximately one-half. This fact is known as the 10°C rule (Layadi et al. 2015).

5.2.2. *Probabilistic Arrhenius model*

5.2.2.1. *Statistic on E_a parameter*

The activation energy (E_a) dramatically impacts the acceleration factor in the Arrhenius equation [5.4].

If we impose a T equal to 80°C and T' as 30°C (delta of 50 K), Figure 5.5 demonstrates the point expressed previously.

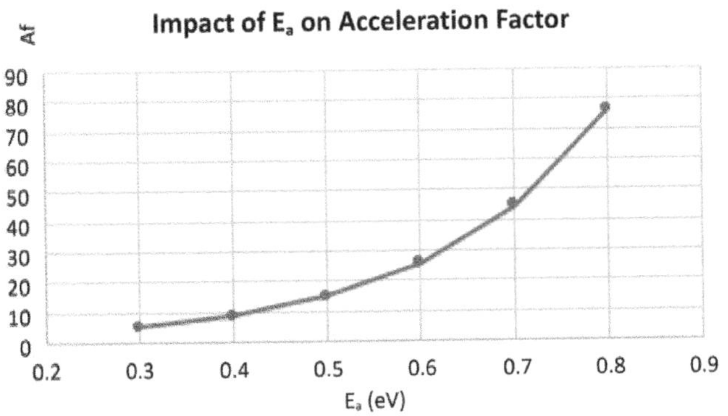

Figure 5.5. *E_a versus acceleration factor*

If the real E_a value of the composite material is 0.6 eV for a specific process, then the real acceleration factor (A_f) is 25.8. In other words, an increase of 50 K divided by almost 26 the lifetime. Nevertheless, if a wrong assumption is used, for instance, 0.79 eV is taken in the Arrhenius law, the A_f amounts 72. It underestimates the lifetime under stress of the product by a factor of 2.8. On the contrary, if 0.4 eV is taken, the A_f amounts 8.7. It overestimates the lifetime and over stress the product by a factor of 3. Hence, the choice of E_a is very important and based on a good knowledge of the material and the mechanism observed in the climatic test.

Even though the E_a is relatively well chosen, an intrinsic variability of the composite material exists. This cannot be avoided. What will be the impact of a small variation on the E_a? Figure 2.2 shows the impact.

If the real E_a is 0.6 eV, a small variation of +/- 0.04 eV gives a variation of A_f from 21 to 32 (real A_f is 26). In other words, 7% of the variation on E_a introduces 20% of the variation on A_f, as shown in Figure 5.6.

Clearly, a very small variation can introduce a large impact on the acceleration factor. Let us take the example in Figure 5.2.

The point demonstrates the utility of integrating a variability of E_a directly in the Arrhenius equation [5.5].

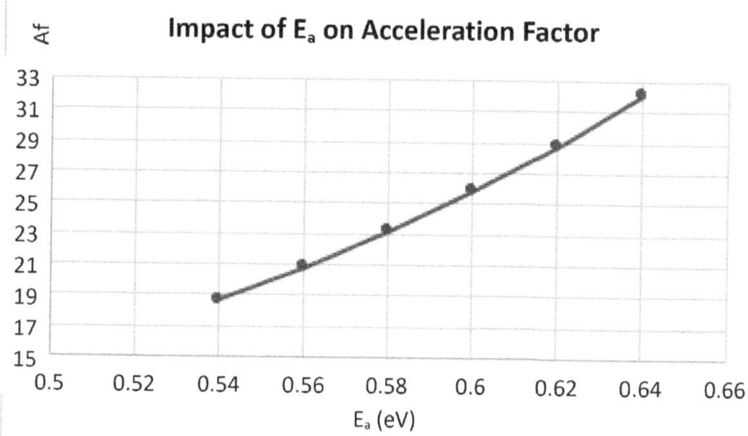

Figure 5.6. *Small variation of E_a versus A_f – Origin Lab software*

Let us apply the Arrhenius equation in an accelerated climatic test set-up for an automotive usage. Equation [5.5] can be written differently as:

$$A_f = \exp\left[-\frac{E_a}{k}\left(\frac{1}{T_b} - \frac{1}{T_u}\right)\right] \qquad [5.5]$$

with T_b being the temperature in the test bench (K) and T_u being the temperature in public road usage (K).

T_e is the equivalent temperature defined by:

$$T_e = \frac{T_u - T_b}{T_u . T_b} \qquad [5.6]$$

In order to simplify the relation [5.4], W is:

$$W = -\frac{T_e}{k} \qquad [5.7]$$

Thus, we have:

$$A_f = \exp(E_a . W) \qquad [5.8]$$

A logarithmic transformation gives:

$$Log\left(A_f\right) = E_a . W \qquad [5.9]$$

Suppose E_a is a random variable distributed by a normal law (Pierrat 2001) defined by:

$$\sigma = \Delta_{E_a} \qquad \mu = \overline{E_a} \qquad [5.10]$$

the probability density function of E_a:

$$f\,(E_a) = \frac{1}{\sqrt{2.\pi.\Delta_{E_a}}}.\,exp\left[-\frac{(E_a-\overline{E_a})^2}{2\Delta_{E_a}{}^2}\right] \qquad [5.11]$$

with, in theory: $-\infty < E_a < +\infty$.

Physically, E_a is necessarily positive.

This assumption of equation [5.10] can be coherent if the variation ΔE_a is small as:

$$\frac{\Delta_{E_a}}{\overline{E_a}} < 30\% \qquad [5.12]$$

If E_a is randomly distributed [5.10], then A_f [5.7], weighted by a W constant, is a random variable distributed by a lognormal law defined by the probability density function:

$$f\,(A_f) = \frac{1}{E_a.\Delta E_a.\sqrt{2.\pi}}.\,exp\left[-\frac{(\ln(E_a)-\overline{E_a})^2}{2\Delta E_a{}^2}\right] \qquad [5.13]$$

The most probable value, also called mode (point of global maximum of the probability density function), is:

$$Mode\,(A_f) = \widehat{A_f} = exp[\mu.W - (\sigma.W)^2] \qquad [5.14]$$

This formula can be interpreted as follows:

If $\Delta E_a = 0$, [5.13] is equivalent to [5.7].

If $\Delta E_a > 0$, the acceleration factor A_f decreases. Of course, [5.13] decreases, if W is lower or T_e is high (T_b-T_u increase).

Therefore, the acceleration factor A_f decreases with the uncertainty of E_a. This phenomenon is more visible if the temperature range of exploration is high.

EXAMPLE 5.1.–

Road public usage: T_u = 40°C (313.15 K).

Bench temperature: $T_b = 120°C$ (393.15 K).

$E_a = 0.85$ eV.

The original Arrhenius formula [5.4] gives:

$$A_f = 607.6 \qquad [5.15]$$

If we introduce a variability of $\Delta E_a = 0.1\ eV$ with a coefficient of variation of 12% [5.11], relation [5.13] gives:

$$\widehat{A_f} = 344.1 \qquad [5.16]$$

What is the real value? It is difficult to define the reality between equations [5.14] and [5.15]. Both values are supposed to describe the real world.

We can prove what is the real value based on experiments on real specimen or on warranty returns analysis.

But if [5.15] is considered as closer to the real value, equation [5.14] does not take into account the variability of E_a and underestimates the stress by 57%.

This simple example proves the added value to introduce the variability of E_a in Arrhenius model.

5.2.2.2. *Statistic on T_u (usage) parameter*

Here, the usage means the mission profile of the product, considering all stress conditions in real life.

Equation [5.4] can be written as:

$$A_f = \exp\left[-\beta\left(\frac{1}{T_b} - \frac{1}{T_u}\right)\right] \qquad \text{with } \beta = \frac{E_a}{k}$$

$$\alpha = \exp\left(\frac{-\beta}{T_b}\right) \quad \text{then } A_f = \alpha \exp\left[\frac{\beta}{T_u}\right] \qquad [5.17]$$

Normally, T_u is a deterministic value that characterizes the ambient temperature stressing the composite in real usage. However, this temperature changes in a specific bandwidth.

Figure 5.7 illustrates the random variability of temperature measured on vehicle closed to a polymer water tank (inlet radiator).

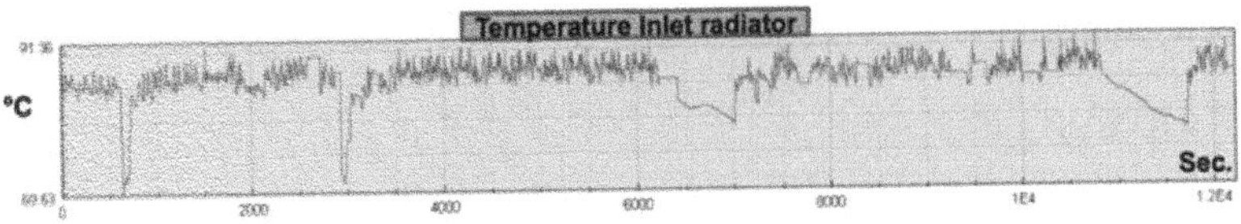

Figure 5.7. *Evolution of temperature inlet radiator with plastic tanks. For a color version of this figure, see www.iste.co.uk/elhami/mechatronics.zip*

We have to consider T_u as a random variable. To simplify the computation, T_u is considered as a random variable distributed by an uniform law (Figure 5.8) limited in this range $[T_a;T_b]$, where $T_a < T_b$ (in K). The random variable T_u has got the same probability in the range $[T_a;T_b]$.

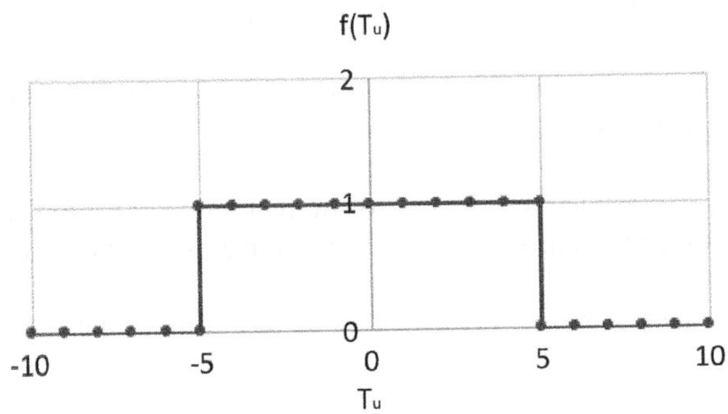

Figure 5.8. *Density function f(Tu)*

$$f_{(T_u)} = \frac{1}{\Delta T} \qquad (17)$$

with $\Delta T = T_b - T_a$ (K).

Equation [5.16] becomes $\widehat{A_f} = \alpha \exp\left[\frac{\beta}{\overline{T_u}}\right]$ (18), with $\overline{T_u}$ being the statistical average of the usage temperature. How to calculate $\overline{T_u}$?

It can be deduced to use a simple arithmetic mean as follows:

$$\overline{T_u} = \frac{T_b - T_a}{2}$$

This approach introduces an important bias of over estimation on the acceleration factor.

Based on the lifetime relationship [5.4], we must solve:

$$\bar{L} = A \exp\left(\frac{\beta}{\overline{T_u}}\right)$$

equivalent to $\frac{\bar{L}}{A} = \bar{\gamma} = \exp\left(\frac{\beta}{\overline{T_u}}\right)$

$\bar{\gamma}$ represents the statistical average of life for a composite at the statistical usage temperature $\overline{T_u}$ moving between T_a and T_b.

Based on equation [5.18], $\overline{T_u}$ can be deduced as $\overline{T_u} = \frac{\beta}{\ln(\bar{\gamma})}$ [5.18]

By solving equation [5.18], the integral defined on $[T_a;T_b]$ becomes:

$$\bar{\gamma} = \int_{T_a}^{T_b} \gamma . f(T_u) . dT_u$$

$$\bar{\gamma} = \int_{T_a}^{T_b} \exp\left(\frac{\beta}{T_u}\right) . \frac{1}{\Delta T} . dT_u$$

with the formula $\int \exp\left(\frac{a}{z}\right) dz = z . e^{\frac{a}{z}} - a . E_i\left(\frac{a}{z}\right)$

$$\bar{\gamma} = \frac{1}{\Delta T}\left[T_b . \exp\left(\frac{\beta}{T_b}\right) - T_a . \exp\left(\frac{\beta}{T_a}\right) - \beta . E_i\left(\frac{\beta}{T_h}\right) + \beta . E_i Ei\left(\frac{\beta}{T_a}\right)\right] \quad [5.19]$$

E_i is the exponential integral defined in Figure 5.9.

E_i can be estimated by a Taylor series from the asymptotic expansion at z>>0 (Abramowitz and Stegun 1968)

$$E_i \approx e^z . \left[\frac{1}{z} + \left(\frac{1}{z}\right)^2 + \frac{2}{z^3} + \frac{6}{z^4} + \frac{24}{z^5} + \frac{120}{z^6} + o\left\{\left(\frac{1}{z}\right)^7\right\}\right]$$

The third order obtains a reasonable error (graph), less than 2% (Figure 5.10) for almost all composite applications [$E_a \in (0.4 - 2.3$ eV) and temperature range (0°C – 250°C)]

$$\bar{\gamma} = \frac{1}{\Delta T}\left[\exp\left(\frac{\beta}{T_a}\right)\left\{\frac{T_a^2}{\beta} + \frac{2.T_a^3}{\beta^2}\right\} - \exp\left(\frac{\beta}{T_b}\right)\left\{\frac{T_b^2}{\beta} + \frac{2.T_b^3}{\beta^2}\right\}\right]$$

If equation [5.22] is substituted into equation [5.21], T_u can be analytically defined.

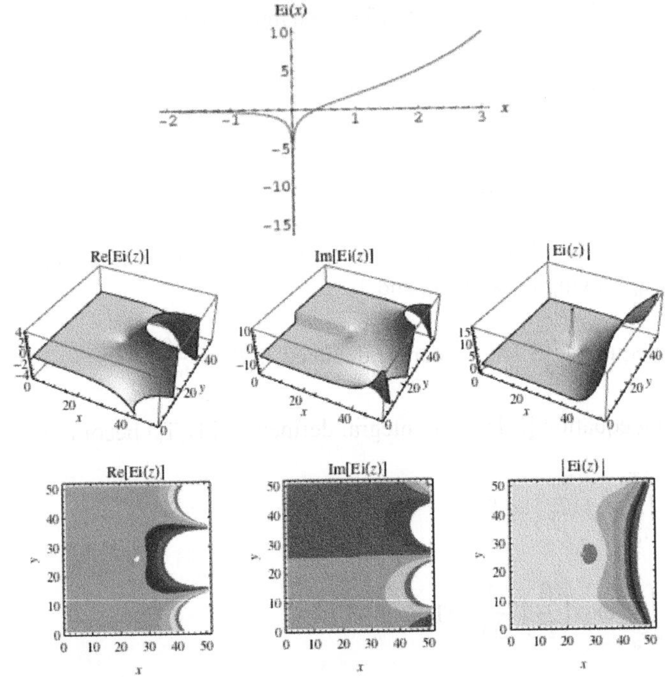

Figure 5.9. *Integral exponential (MathoWorld). For a color version of this figure, see www.iste.co.uk/elhami/mechatronics.zip*

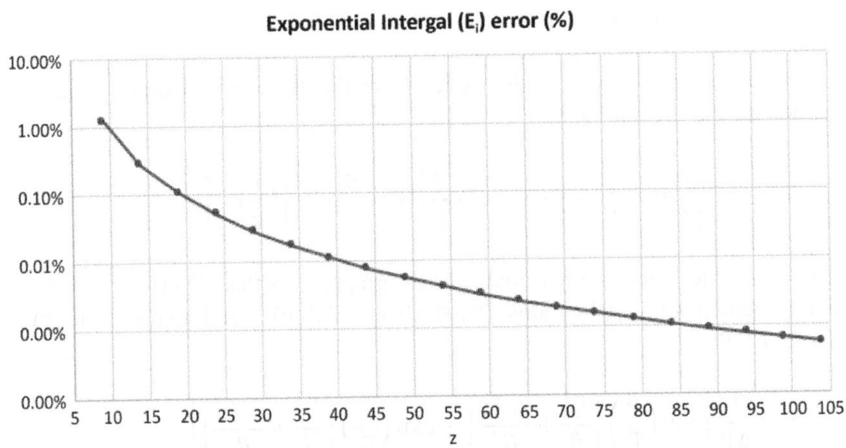

Figure 5.10. *Exponential integral (Ei)*

EXAMPLE 5.2.–

Bench temperature: T_b = 120°C (393.15 K).

E_a = 0.85 eV (supposed fix).

Road public usage: T_u can evaluate with the same probability in this range $[T_a;T_b]$ = [20°C; 60°C] (293.15 K; 333.15 K).

The statistical average value $\overline{T_u}$ is 34°C (306.73 K). This value is lower than the simple arithmetic which gives 40°C as in Example 5.1.

The statistical average of A_f is:

$$\widehat{A_f} = 1{,}174.9 \quad [5.20]$$

The probabilistic Arrhenius model on T_u parameters allows us to reduce the test time by almost a factor of 2 (ratio (23)/(14)).

This formulation is slightly (1.06%) more conservative (Pierrat and Feidt 2013).

5.2.2.3. *Statistic on T_b (bench temperature) parameter*

Based on the same scientific approach developed previously, it is possible to appreciate the bench temperature variability. This variability can be the picture of the control sensor uncertainty in the climatic chamber (usually less than 2 K):

$$A_f = \exp\left[\beta'\left(\frac{1}{T_b} - \frac{1}{T_u}\right)\right] \qquad \text{with } \beta' = -\frac{E_a}{k}$$

$$\alpha' = \exp\left(\frac{-\beta'}{T_u}\right), \quad \text{then } \widehat{A_f} = \alpha' \exp\left[\frac{\beta'}{T_b}\right] \quad [5.21]$$

T_b is a random variable defined by a uniform distribution, equiprobability in the area $[T_a'; T_b']$, where $T_a' < T_b'$:

$$\overline{\gamma}' = \frac{1}{\Delta T'.\beta'}\left[T_{a'}^2.\exp\left(\frac{\beta'}{T_{a'}}\right)\left\{1 + \frac{2.T_{a'}}{\beta'}\right\} - T_{b'}^2.\exp\left(\frac{\beta'}{T_{b'}}\right)\left\{1 + \frac{2.T_{b'}}{\beta'}\right\}\right] \quad [5.22]$$

$$\overline{T_b}' \text{ can be deduced as } \overline{T_b}' = \frac{\beta'}{\ln(\gamma')} \quad [5.23]$$

EXAMPLE 5.3.–

Road public usage: T_u = 40°C (313.15 K).

$E_a = 0.85$ eV.

Bench temperature: $T_b = 120°C$ (393.15 K) with a small variation of 0.1 K (with a Pt100 sensor to control the climatic chamber for instance).

The statistical average of A_f is:

$$\widehat{A_f} = 601.8 \qquad [5.24]$$

The uncertainty of the bench temperature (climatic chamber) reduces the acceleration factor (ratio (27)/(14)) about 1%.

a) Combination of the three probabilistic models:

The Arrhenius model defined by the acceleration factor A_f [5.4]:

$$A_f = \exp\left[-\frac{E_a}{k}\left(\frac{1}{T_b} - \frac{1}{T_u}\right)\right]$$

uses three different parameters E_a, T_b, T_u, which can be evaluated by probabilistic models [5.13], [5.24] and [5.18] respectively. These statistical variables are independent of one another. Consequently, the acceleration factor is the sum of all variability. In other words:

$$\widehat{A_f} = \sum_{i=1}^{k} \overline{A_{fi}} - kA_f \ (24)$$

or

$$\widehat{A_f} = \overline{A_f(T_u)} + \overline{A_f(T_b)} + \overline{A_f(E_a)} - 3.A_f \qquad [5.25]$$

with A_f being defined by a constant Arrhenius model (4); $\overline{A_f(T_u)}$ being defined by a variability of the usage temperature [5.18]; $\overline{A_f(T_b)}$ being defined by a variability of the bench temperature [5.24]; $\overline{A_f(T_{Ea})}$ being defined by a variability of E_a [5.13].

Considering Examples 5.1–5.3:

$$\widehat{A_f} = 298$$

This means, if the road public usage is 15.000 hours (cumulative usage time for 15 years), the constant Arrhenius model defines an equivalent test time of 25 hours. The new Arrhenius probabilistic models assesses an equivalent test time of 50 hours.

Figure 5.11 describes the influence of all statistical dispersions of the three parameters.

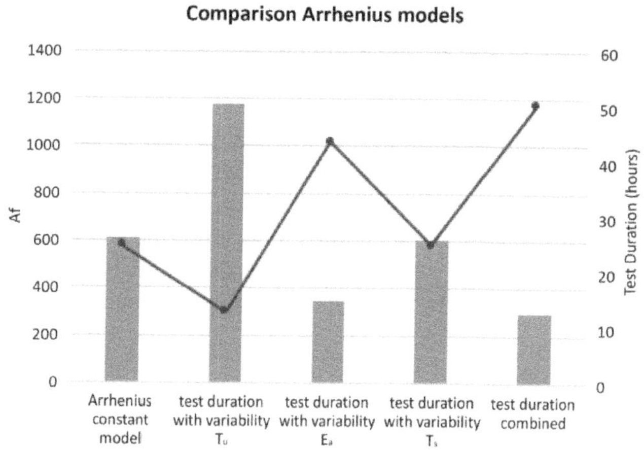

Figure 5.11. *Comparison of Arrhenius models. For a color version of this figure, see www.iste.co.uk/elhami/mechatronics.zip*

5.3. Experimental case

5.3.1. *Assumptions*

According to a German carmaker, the specimen is shown to undergo the temperature distribution presented in Table 5.1.

Temperature (°C)	Percentile (%)
-40	6
23	20
40	65
75	8
85	1

Table 5.1. *Distribution of the temperature*

The maximum temperature that can be reached by the specimen is 85°C.

This mission profile is defined by one temperature associated with a percentile of the lifetime of the system of approximately 15 years, 300,000 km and 8,000 hours of working.

Note that 8,000 hours in 15 years is equivalent to 533 hours of working per year. According to the state of the art about the driving pattern in Europe, this amount corresponds to 70% of driver severity.

Table 5.1 gives some extract of the mission profile based on some points extracted from a statistical distribution. Indeed, the temperature distribution is not just 6 points, but a continuous mathematical function is represented in Figure 5.12(b). Based on this fact, it is more realistic to write the mission profile not by some points but with a "band" of temperature, as shown in Figure 5.12(a). In this example, the "band" is defined by the temperature plus or minus 5 K.

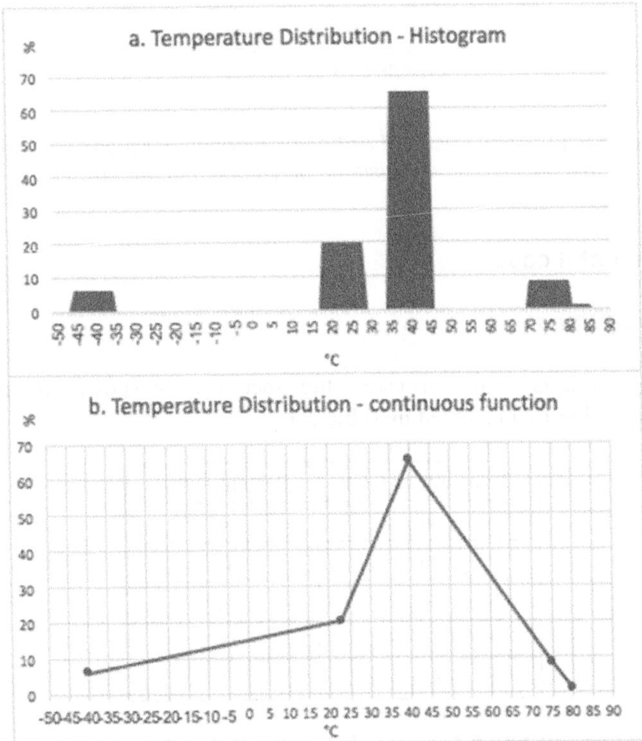

Figure 5.12. *Temperature distribution. For a color version of this figure, see www.iste.co.uk/elhami/mechatronics.zip*

Therefore, in the Arrhenius model, the usage temperature T_u, can be defined as in Table 5.2.

Temperature band (°C)	Percentile (%)
(-45;-35)	6
(18;28)	20
(35–45)	65
(70;80)	8
(75;85)	1

Table 5.2. *Distribution of the temperature by band*

This mission profile assumes that worldwide usage is covered with the objective of 95% of reliability, a level of confidence of 80% at 15 years and 300,000 km. The warranty period is three years.

In this concrete case, the product of interest it is an automotive sensor which is placed outside of the car. The product contains a sealing and it can be assessed a value for E_a of 0.64 eV +/- 0.15 eV for a thermal-oxidation failure mechanism of sealing material in such an automotive system (Kömmling et al. 2017). During product validation, the sensor is exposed to different tests, such as the thermal life test. To assess functionality and tightness, the product is continuously monitored during the tests, and a tightness check is carried out at defined intervals by using a leakage test with 0.2 bar pressure overload.

Figure 5.13. *Product placed in a climate chamber. For a color version of this figure, see www.iste.co.uk/elhami/mechatronics.zip*

Finally, the temperature of the climate chamber is controlled with +/- 2 K at the maximum temperature of the tested system +85°C.

Figure 5.13 shows several products placed inside a climate chamber.

5.3.2. *Results*

According to the carmaker, the duration for a thermal life test is defined as 1,656 hours at maximum temperature.

According to the constant Arrhenius model, the acceleration factor is 11.5 and an equivalent test time of 697.4 hours to cover 15 years, 300,000 km or 8,000 hours of working is obtained.

Based on the probabilistic Arrhenius model, the acceleration factor is 8.8 or 908.1 hours to represent the total lifetime.

Figure 5.14 shows the impact of all the parameters.

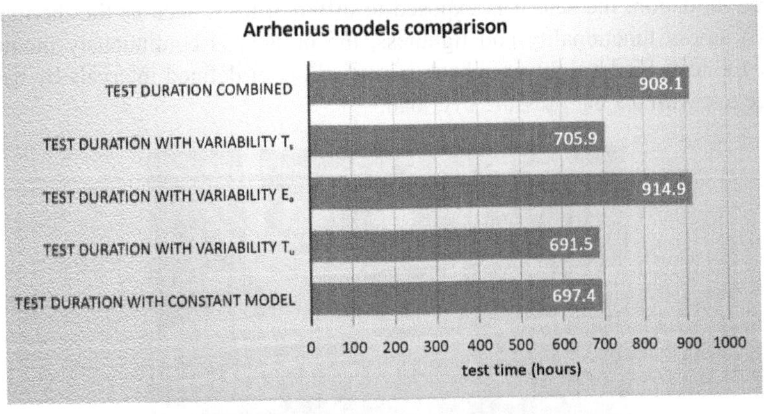

Figure 5.14. *Comparison of Arrhenius models. For a color version of this figure, see www.iste.co.uk/elhami/mechatronics.zip*

5.3.3. *Exploration*

Following Figure 5.14, there is a significant gap between the constant Arrhenius model (close to 700 hours of tests) and the probabilistic Arrhenius model (more than 900 hours).

What should the risk be if the constant Arrhenius model is applied?

Based on the binomial law (equation [5.1]), the test can be fixed as follows:

$$T = t\,.\left[\frac{Ln\,(\alpha)}{n.Ln(R_t)}\right]^{\frac{1}{\beta}} \qquad [5.26]$$

with:

t: the initial test time;

n: the number of specimens tested;

β: the shape parameter for a two-parameter Weibull distribution;

C: the level of confidence;

α = 1-C;

R: the reliability target.

In this example, it is assumed that:

T: 908 h, n = 6, α = 20%, R = 95%, β = 2 (degradation phenomenon). In other words, it is assumed that the probabilistic Arrhenius model covers the reliability target of the customer and is equivalent damage of the total lifetime.

In this case, if 697 hours of tests (constant Arrhenius model) are applied, only 91.67% of reliability (based on equation [5.29]) is achieved.

Clearly, even though the customer target is not achieved, it can be "negligible" to apply a constant Arrhenius model instead of the probabilistic model.

What should be the cost impact on the warranty period of three years?

To assess the financial risk, a two-parameter Weibull distribution can be used as:

$$F(t) = 1 - Exp\left[-\left(\frac{t}{\eta}\right)^{\beta}\right] \qquad [5.27]$$

with:

η: the scale parameter;

F: the probability of failure;

β: the shape parameter.

With the reliability target (95%), it is possible to calculate the scale parameter with $\beta = 2$ (degradation phenomenon) at $t = 15$ years. Thus, $\eta = 66$ years is obtained.

Now, the probability of failure in the warranty period of three years is:

$$F\ (3\ years) = 1 - Exp\left[-\left(\frac{3}{66}\right)^2\right] = 0.21\%\ of\ failure$$

Therefore, 0.21% of failure in three years is the expected warranty cost with a reliable design demonstrated with the probabilistic Arrhenius model.

With the constant Arrhenius model, the reliability demonstration is only 91.7% at 15 years. As discussed previously, the scale parameter is defined with $\eta = 51$ years. In this case, the probability of failure in the warranty period can be assessed as follows:

$$F\ (3\ years) = 1 - Exp\left[-\left(\frac{3}{51}\right)^2\right] = 0.35\%\ of\ failure$$

Therefore, 0.35% of failure in three years is the expected warranty cost demonstrated with the constant Arrhenius model.

In other words, there would be almost twice (1.7 times) more warranty costs if the reliability demonstration is limited with the constant Arrhenius model for this specific failure mechanism. This clearly illustrates the importance of using the probabilistic Arrhenius model.

5.3.4. *Results and discussions*

In this section, we present the results by pointing out on our own contributions to the field.

	Sine wave	Square wave
Crest factor	1.414	1
RMS value	0.707	1

Table 5.3. *RMS and crest factor for sine and square waves*

5.4. Conclusion and outlook

The Arrhenius relationship has been widely used for decades. Some recent applications are in, for example, medical devices (Jiang et al. 2003), lithium ion cells (Broussely et al. 2001), petroleum-based ferrofluid (Segal et al. 1999) and motor insulation systems (Oraee 2000). However, it should be noted that the Arrhenius relationship is not universally applicable to all cases where temperature is an accelerating stress. Some examples are reported in Gillen et al. (2005b) on a commercial chloroprene rubber cable jacketing material, as well as in Dimaria and Stathis (1999) on an ultrathin silicon dioxide film. Therefore, this proposed probabilistic model is developed to cover reliability tests based on Arrhenius for mechatronic products where many different materials are used with their characteristic activation energies related to aging processes and climate load. In the first experimental phase, the usage of the model can be proved; further tests have to be carried out to underline this. It is important to verify the adequacy of the model by using test data.

5.5. References

Abramowitz, M. and Stegun, I.A. (1968). *Handbook of Mathematical Functions*. Dover Publications, New York.

Aitchison, J. and Brown, J.A.C. (1957). *The Lognormal Distribution*. Cambridge University Press, Cambridge.

Arrhenius, S. (1889). Über die Reaktion-geschwindigkeit bei der inversion vor Rhorzucker durch Sauren. *Zeitschrift für Physikalische Chemie*, volume 4U, issue: 1. DOI: 10.1515/zpch-1889-0416.

Bernstein, R. (2010). Nylon 6.6 accelerating aging studies 2: Long-term thermal-oxidative and hydrolysis results. *Polymer Degradation and Sustainability*, 95(9), 1471–1479, DOI:10.1016/j.polymdegradstab.2010.06.018.

Bertsche, B. (2008). *Reliability Automotive and Mechanical Engineering*. Springer, Berlin, Heidelberg.

Broussely, M., Herreyre, S., Biensan, P., Kasztejna, P., Nechev, K., Staniewicz, R.J. (2001). Aging mechanism in Li ion cells and calendar life predictions. *Journal of Power Sources*, 97–98, 13–21.

Charette, R.N. (2009). This car runs on code. *IEEE Spectrum* [Online]. Available at: https://spectrum.ieee.org/this-car-runs-on-code.

Delaux, D., El Hami, A., Grzeskowiak, H. (2017a). *Reliability of High-Power Mechatronics Systems 1*. ISTE Press, London, and Elsevier, Oxford.

Delaux, D., El Hami, A., Grzeskowiak, H. (2017b). *Reliability of High-Power Mechatronics Systems 2*. ISTE Press, London, and Elsevier, Oxford.

Demir, A.D. and Cronin, K. (2004). Modelling the kinetics of textural changes in hazelnuts during roasting. *Journal Simulation Modelling Practice and Theory* [Online]. Available at: https://doi.org/10.1016/j.simpat.2003.11.007.

Dimaria, D.J. and Stathis, J.H. (1999). Non-Arrhenius temperature dependence of reliability in ultrathin silicon dioxide films. *Applied Physics Letters*, 74(12), 1752–1754.

Eriksson, A., Albertsson, A.-C., Boydell, P., Månson, J.A.E. (2004). Durability of inplant recycled glass fiber reinforced polyamide 66. *Polymer Engineering and Science*, 38 [Online]. Available at: https://doi.org/10.1002/pen.10196.

Gibrat, R. (1930). Une loi des répartitions économiques : l'effet proportionnel. *Bull. Stat. Gen. Fr.*, 19.

Gillen, K.T., Bernstein, R., Celina, M. (2005a). Non-Arrhenius behavior for oxidative degradation of chlorosulfonated polyethylene materials. *Polymer Degradation and Stability*, 87(2), 335–346. DOI: 10.1016/j.polymdegradstab.2004.09.004.

Gillen, K.T., Bernstein, R., Derzon, D.K. (2005b). Evidence of non-Arrhenius behavior from laboratory aging and 24-year field aging of polychloroprene rubber materials. *Polymer Degradation and Stability*, 87(1), 57–67.

Gillen, K.T., Bernstein, R., Clough, R.L., Celina, M. (2006). Lifetime predictions for semi-crystalline cable insulation materials I: Mechanical properties and oxygen consumption measurements on EPR materials, 2006 [Online]. Available at: https://doi.org/10.1016/j.polymdegradstab.2006.01.009.

Illing, T., Schoßig, M., Bierögel, C., Langer, B., Grellmann, W. (2017). Hygrothermal aging of injection-moulded PA6/GF materials considering automotive requirements. *Grellmann, Wolfgang, and Beate Langer. Deformation and Fracture Behaviour of Polymer Materials*, 405–419.

Jiang, G., Purnell, K., Mobley, P., Shulman, J. (2003). Accelerated life tests and in-vivo test of 3Y-TZP ceramics. *Proc. Materials and Processes for Medical Devices Conference*, ASM International, 477–482.

Kömmling, A., Jaunich, M., Pourmand, P., Wolff, D., Gedde, U.W. (2017). Influence of ageing on sealability of elastomeric O-rings. *Macromolecular Symposia*, 373(1).

Koopmans, L.H., Owen, D.B., Rosenblatt, J.I. (1964). Confidence intervals for the coefficient of variation for the normal and lognormal distributions. *Biometrika*, 51, 25–32. DOI: 10.1093/biomet/51.1-2.25.

Lannoy, A. (2008). *Maîtrise des risques et sûreté de fonctionnement. Repères historiques et méthodologiques*. Tech & Doc, Lavoisier, Paris.

Layadi, T.M., Champenois, G., Mostefai, M., Abbes, D. (2015). Lifetime estimation tool of lead-acid batteries for hybrid power sources design. *Journal Simulation Modelling Practice and Theory* [Online]. Available at: https://doi-org.ezproxy.normandie-univ.fr/10.1016/j.simpat.2015.03.001.

MathWorld (2017). Exponential integral [Online]. Available at: http://mathworld.wolfram.com/ExponentialIntegral.html.

Nosratabadi, S.M., Hooshmand, R.-A., Gholipour, E., Rahimi, S. (2017). Modeling and simulation of long term stochastic assessment in industrial microgrids proficiency considering renewable resources and load growth. *Journal Simulation Modelling Practice and Theory* [Online]. Available at: https://doi-org.ezproxy.normandie-univ.fr/10.1016/j.simpat.2017.03.013.

Oraee, H. (2000). Quantative approach to estimate the life expectancy of motor insulation systems. *IEEE Transactions on Dielectrics and Electrical Insulation*, 7(6), 790–796.

Pierrat, L. (2001). Modèle d'Arrhenius : mieux estimer le facteur d'accélération thermique. *Journal Essai Industriel*, March, 16, 28–30.

Pierrat, L. and Feidt, M. (2013). Introduction d'une variation aléatoire de témperature dans le modèle de dégradation thermodynamique d'Arrhenius. *Société française de thermique*, Congress.

Plastic Makes it Possible (2015). Automakers turn to plastics for lighter vehicles, increased fuel efficiency [Online]. Available at: https://www.plasticsmakeitpossible.com/whats-new-cool/automotive/fuelefficiency/automakers-turn-to-plastics-for-lighter-vehicles-increased-fuel-efficiency/[Accessed 21 October 2017].

Prabhu, N.U. (1965). *Stochastic Processes: Basic Theory and Its Applications*. Macmillan, New York.

Rodat, D., Guibert, F., Dominguez, N., Calmon, P. (2018). Introduction of physical knowledge in kriging-based meta-T modelling approaches applied to non-destructive testing simulations. *Journal Simulation Modelling Practice and Theory* [Online]. Available at: https://doi-org.ezproxy.normandie-univ.fr/10.1016/j.simpat.2010.06.002.

Segal, V., Nattrass, D., Raj, K., Leonard, D. (1999). Accelerated thermal aging of petroleum-based ferrofluids. *Journal of Magnetism and Magnetic Materials*, 201, 70–72.

Toyota (n.d.). Estimation du nombre de composant dans un véhicule [Online]. Available at: http://www.toyota.co.jp/en/kids/faq/d/01/04/.

Verma, A.K., Ajit, S., Karanki, D.R. (2010). *Reliability and Safety Engineering*. Springer, Berlin, Heidelberg.

Weibull, W. (1951). A statistical distribution function of wide applicability. *Journal of Applied Mechanics, Transactions ASME*, 18(3), 293–297.

Wikimedia Commons (2020a). File:Unidentified rural letter carrier with modified Model-T Ford.jpg [Online]. Available at: https://commons.wikimedia.org/wiki/File:Unidentified_rural_letter_carrier_with_modified_Model-T_Ford.jpg.

Wikimedia Commons (2020b). File:2016-03-01 Geneva Motor Show 0999.JPG [Online]. Available at: https://commons.wikimedia.org/wiki/File:2016-03-01_Geneva_Motor_Show_0999.JPG?uselang=fr.

Wikimedia Commons (2022). File:Model T on a New Road (8113428270).jpg [Online]. Available at: https://commons.wikimedia.org/wiki/File:Model_T_on_a_New_Road_(8113428270).jpg.

Yadolah, D. (1993). *Statistique : dictionnaire encyclopédique*. Dunod, Paris.

Yang, G. (2007). *Lifecycle, Reliability Engineering*. John Wiley & Sons, New York.

List of Authors

Alain BENSOUSSAN
Consultant
France

Joseph B. BERNSTEIN
Ariel University
Israel

Alain BRAVAIX
STMicroelectronics
Grenoble
France

Felician CAMPEAN
Bradford University
UK

David DELAUX
Valeo Thermal System
Paris
France

Abdelkhalak EL HAMI
INSA
University of Rouen
France

Henri GRZESKOWIAK
HGC
France

Ed HENSHALL
Bradford University
UK

Thomas ILLING
Valeo Comfort Driving Assistance
Bietigheim-Bissingen
Germany

Index

A, B, C, D

E, F, G, H

L, M, P

R, S, U, V, W

Other titles from

ISTE

in

Mechanical Engineering and Solid Mechanics

2022

BAYLE Franck
Product Maturity 1: Theoretical Principles and Industrial Applications
(Reliability of Multiphysical Systems Set – Volume 12)
Product Maturity 2: Principles and Illustrations
(Reliability of Multiphysical Systems Set – Volume 13)

LANNOY André
Reliability of Nuclear Power Plants: Methods, Data and Applications
(Reliability of Multiphysical Systems Set – Volume 14)

LEDOUX Michel, EL HAMI Abdelkhalak
Heat Transfer 3: Convection, Fundamentals and Monophasic Flows
(Mathematical and Mechanical Engineering Set – Volume 11)

2021

CHALLAMEL Noël, KAPLUNOV Julius, TAKEWAKI Izuru
Modern Trends in Structural and Solid Mechanics 1: Static and Stability
Modern Trends in Structural and Solid Mechanics 2: Vibrations
Modern Trends in Structural and Solid Mechanics 3: Non-deterministic Mechanics

DAHOO Pierre Richard, POUGNET Philippe, EL HAMI Abdelkhalak
Applications and Metrology at Nanometer Scale 1: Smart Materials, Electromagnetic Waves and Uncertainties
(Reliability of Multiphysical Systems Set – Volume 9)
Applications and Metrology at Nanometer Scale 2: Measurement Systems, Quantum Engineering and RBDO Method
(Reliability of Multiphysical Systems Set – Volume 10)

LEDOUX Michel, EL HAMI Abdelkhalak
Heat Transfer 1: Conduction
(Mathematical and Mechanical Engineering Set – Volume 9)
Heat Transfer 2: Radiative Transfer
(Mathematical and Mechanical Engineering Set – Volume 10)

PLANCHETTE Guy
Cindynics, The Science of Danger: A Wake-up Call
(Reliability of Multiphysical Systems Set – Volume 11)

2020

SALENÇON Jean
Elastoplastic Modeling

2019

BAYLE Franck
Reliability of Maintained Systems Subjected to Wear Failure Mechanisms: Theory and Applications
(Reliability of Multiphysical Systems Set – Volume 8)

BEN KAHLA Rabeb, BARKAOUI Abdelwahed, MERZOUKI Tarek
Finite Element Method and Medical Imaging Techniques in Bone Biomechanics
(Mathematical and Mechanical Engineering Set – Volume 8)

IONESCU Ioan R., QUEYREAU Sylvain, PICU Catalin R., SALMAN Oguz Umut
Mechanics and Physics of Solids at Micro- and Nano-Scales

LE VAN Anh, BOUZIDI Rabah
Lagrangian Mechanics: An Advanced Analytical Approach

MICHELITSCH Thomas, PÉREZ RIASCOS Alejandro, COLLET Bernard, NOWAKOWSKI Andrzej, NICOLLEAU Franck
Fractional Dynamics on Networks and Lattices

SALENÇON Jean
Viscoelastic Modeling for Structural Analysis

VÉNIZÉLOS Georges, EL HAMI Abdelkhalak
Movement Equations 5: Dynamics of a Set of Solids
(Non-deformable Solid Mechanics Set – Volume 5)

2018

BOREL Michel, VÉNIZÉLOS Georges
Movement Equations 4: Equilibriums and Small Movements
(Non-deformable Solid Mechanics Set – Volume 4)

FROSSARD Etienne
Granular Geomaterials Dissipative Mechanics: Theory and Applications in Civil Engineering

RADI Bouchaib, EL HAMI Abdelkhalak
Advanced Numerical Methods with Matlab® 1: Function Approximation and System Resolution (Mathematical and Mechanical Engineering SET – Volume 6)
Advanced Numerical Methods with Matlab® 2: Resolution of Nonlinear, Differential and Partial Differential Equations
(Mathematical and Mechanical Engineering SET – Volume 7)

SALENÇON Jean
Virtual Work Approach to Mechanical Modeling

2017

BOREL Michel, VÉNIZÉLOS Georges
Movement Equations 2: Mathematical and Methodological Supplements
(Non-deformable Solid Mechanics Set – Volume 2)
Movement Equations 3: Dynamics and Fundamental Principle
(Non-deformable Solid Mechanics Set – Volume 3)

BOUVET Christophe
Mechanics of Aeronautical Solids, Materials and Structures
Mechanics of Aeronautical Composite Materials

BRANCHERIE Delphine, FEISSEL Pierre, BOUVIER Salima, IBRAHIMBEGOVIC Adnan
From Microstructure Investigations to Multiscale Modeling: Bridging the Gap

CHEBEL-MORELLO Brigitte, NICOD Jean-Marc, VARNIER Christophe
From Prognostics and Health Systems Management to Predictive Maintenance 2: Knowledge, Traceability and Decision
(Reliability of Multiphysical Systems Set – Volume 7)

EL HAMI Abdelkhalak, RADI Bouchaib
Dynamics of Large Structures and Inverse Problems
(*Mathematical and Mechanical Engineering Set – Volume 5*)
Fluid-Structure Interactions and Uncertainties: Ansys and Fluent Tools
(Reliability of Multiphysical Systems Set – Volume 6)

KHARMANDA Ghias, EL HAMI Abdelkhalak
Biomechanics: Optimization, Uncertainties and Reliability
(Reliability of Multiphysical Systems Set – Volume 5)

LEDOUX Michel, EL HAMI Abdelkhalak
Fluid Mechanics: Analytical Methods
(Mathematical and Mechanical Engineering Set – Volume 3)
Compressible Flow Propulsion and Digital Approaches in Fluid Mechanics
(Mathematical and Mechanical Engineering Set – Volume 4)

MORI Yvon
Mechanical Vibrations: Applications to Equipment

2016

BOREL Michel, VÉNIZÉLOS Georges
Movement Equations 1: Location, Kinematics and Kinetics
(Non-deformable Solid Mechanics Set – Volume 1)

BOYARD Nicolas
Heat Transfer in Polymer Composite Materials

CARDON Alain, ITMI Mhamed
New Autonomous Systems
(Reliability of Multiphysical Systems Set – Volume 1)

DAHOO Pierre Richard, POUGNET Philippe, EL HAMI Abdelkhalak
Nanometer-scale Defect Detection Using Polarized Light
(Reliability of Multiphysical Systems Set – Volume 2)

DE SAXCÉ Géry, VALLÉE Claude
Galilean Mechanics and Thermodynamics of Continua

DORMIEUX Luc, KONDO Djimédo
Micromechanics of Fracture and Damage
(Micromechanics Set – Volume 1)

EL HAMI Abdelkhalak, RADI Bouchaib
Stochastic Dynamics of Structures
(Mathematical and Mechanical Engineering Set – Volume 2)

GOURIVEAU Rafael, MEDJAHER Kamal, ZERHOUNI Noureddine
From Prognostics and Health Systems Management to Predictive Maintenance 1: Monitoring and Prognostics
(Reliability of Multiphysical Systems Set – Volume 4)

KHARMANDA Ghias, EL HAMI Abdelkhalak
Reliability in Biomechanics
(Reliability of Multiphysical Systems Set – Volume 3)

MOLIMARD Jérôme
Experimental Mechanics of Solids and Structures

RADI Bouchaib, EL HAMI Abdelkhalak
Material Forming Processes: Simulation, Drawing, Hydroforming and Additive Manufacturing (Mathematical and Mechanical Engineering Set – Volume 1)

2015

KARLIČIĆ Danilo, MURMU Tony, ADHIKARI Sondipon, MCCARTHY Michael
Non-local Structural Mechanics

SAB Karam, LEBÉE Arthur
Homogenization of Heterogeneous Thin and Thick Plates

2014

ATANACKOVIC M. Teodor, PILIPOVIC Stevan, STANKOVIC Bogoljub, ZORICA Dusan
Fractional Calculus with Applications in Mechanics: Vibrations and Diffusion Processes
Fractional Calculus with Applications in Mechanics: Wave Propagation, Impact and Variational Principles

CIBLAC Thierry, MOREL Jean-Claude
Sustainable Masonry: Stability and Behavior of Structures

ILANKO Sinniah, MONTERRUBIO Luis E., MOCHIDA Yusuke
The Rayleigh−Ritz Method for Structural Analysis

LALANNE Christian
Mechanical Vibration and Shock Analysis – 5-volume series – 3rd edition
Sinusoidal Vibration – Volume 1
Mechanical Shock – Volume 2
Random Vibration – Volume 3
Fatigue Damage – Volume 4
Specification Development – Volume 5

LEMAIRE Maurice
Uncertainty and Mechanics

2013

ADHIKARI Sondipon
Structural Dynamic Analysis with Generalized Damping Models: Analysis

ADHIKARI Sondipon
Structural Dynamic Analysis with Generalized Damping Models: Identification

BAILLY Patrice
Materials and Structures under Shock and Impact

BASTIEN Jérôme, BERNARDIN Frédéric, LAMARQUE Claude-Henri
Non-smooth Deterministic or Stochastic Discrete Dynamical Systems: Applications to Models with Friction or Impact

EL HAMI Abdelkhalak, RADI Bouchaib
Uncertainty and Optimization in Structural Mechanics

KIRILLOV Oleg N., PELINOVSKY Dmitry E.
Nonlinear Physical Systems: Spectral Analysis, Stability and Bifurcations

LUONGO Angelo, ZULLI Daniele
Mathematical Models of Beams and Cables

SALENÇON Jean
Yield Design

2012

DAVIM J. Paulo
Mechanical Engineering Education

DUPEUX Michel, BRACCINI Muriel
Mechanics of Solid Interfaces

ELISHAKOFF Isaac *et al.*
Carbon Nanotubes and Nanosensors: Vibration, Buckling and Ballistic Impact

GRÉDIAC Michel, HILD François
Full-Field Measurements and Identification in Solid Mechanics

GROUS Ammar
Fracture Mechanics - 3-volume series
Analysis of Reliability and Quality Control – Volume 1
Applied Reliability – Volume 2
Applied Quality Control – Volume 3

RECHO Naman
Fracture Mechanics and Crack Growth

2011

KRYSINSKI Tomasz, MALBURET François
Mechanical Instability

SOUSTELLE Michel
An Introduction to Chemical Kinetics

2010

BREITKOPF Piotr, FILOMENO COELHO Rajan
Multidisciplinary Design Optimization in Computational Mechanics

DAVIM J. Paulo
Biotribolgy

PAULTRE Patrick
Dynamics of Structures

SOUSTELLE Michel
Handbook of Heterogenous Kinetics

2009

BERLIOZ Alain, TROMPETTE Philippe
Solid Mechanics using the Finite Element Method

LEMAIRE Maurice
Structural Reliability

2007

GIRARD Alain, ROY Nicolas
Structural Dynamics in Industry

GUINEBRETIÈRE René
X-ray Diffraction by Polycrystalline Materials

KRYSINSKI Tomasz, MALBURET François
Mechanical Vibrations

KUNDU Tribikram
Advanced Ultrasonic Methods for Material and Structure Inspection

SIH George C. *et al.*
Particle and Continuum Aspects of Mesomechanics

Printed and bound by CPI Group (UK) Ltd, Croydon, CR0 4YY

16/09/2026

14963081-0006